Modelling and Simulation of Power Generation Plants

Other titles published in this Series:

Parallel Processing for Jet Engine Control
Haydn A. Thompson

Iterative Learning Control for Deterministic Systems
Kevin L. Moore

Parallel Processing in Digital Control
D. Fabian Garcia Nocetti and Peter J. Fleming

Intelligent Seam Tracking for Robotic Welding
Nitin Nayak and Asok Ray

Identification of Multivariable Industrial Processes for Simulation, Diagnosis and Control
Yucai Zhu and Ton Backx

Nonlinear Process Control: Applications of Generic Model Control
Edited by Peter L. Lee

Microcomputer-Based Adaptive Control Applied to Thyristor-Driven D-C Motors
Ulrich Keuchel and Richard M. Stephan

Expert Aided Control System Design
Colin Tebbutt

Modeling and Advanced Control for Process Industries Applications to Paper Making Processes
Ming Rao, Qijun Xia and Yiqun Ying

Robust Multivariable Flight Control
Richard J. Adams, James M. Buffington, Andrew G. Sparks and Siva S. Banda

A. W. Ordys, A. W. Pike, M. A. Johnson,
R. M. Katebi and M. J. Grimble

Modelling and Simulation of Power Generation Plants

With 98 Figures

Springer-Verlag
London Berlin Heidelberg New York
Paris Tokyo Hong Kong
Barcelona Budapest

Andrzej Ordys, MScEE, PhD
Reza Katebi, BSc, MSc, PhD, MIEE, CEng, MIEEE
Michael A. Johnson, BSc, DIC, MSc, PhD, FIMA, CMath, MIEEE
Michael J. Grimble, BSc, BA, MSc, PhD, DSc, FIEEE, FIEE, FInstMC, FIMA, CEng, CMath

Industrial Control Centre
University of Strathclyde
50 George Street
Glasgow G1 1QE, Scotland, UK

ISBN-13:978-1-4471-2116-9 e-ISBN-13:978-1-4471-2114-5
DOI: 10.1007/978-1-4471-2114-5

British Library Cataloguing in Publication Data
A catalogue record for this book is available from the British Library

Library of Congress Cataloging-in-Publication Data
A catalog record for this book is available from the Library of Congress

Softcover reprint of the hardcover 1st edition 1994

Typesetting: Camera-ready by authors

69/3830–543210 Printed on acid-free paper

For their love and support

To my wife Anna and my children Szymon and Bartek

Andrzej Ordys

To my wife Niki, my sister Melissa, and to my Mother and Father

Andrew Pike

To Eleanor and Laurence, and my Mother and Father

Michael Johnson

To my wife Haleh, and to my children Raha and Imaon

Reza Katebi

To Wendy, Claire and Andrew and my Mother and Father

Michael Grimble

FOREWORD

Many large-scale processes like refineries or power generation plant are constructed using the multi-vendor system and a main co-ordinating engineering contractor. With such a methodology the key process units are installed complete with local proprietary control systems in place. Re-assessing the so called lower level control loop design or structure is becoming less feasible or desirable. Consequently, future competitive gains in large-scale industrial systems will arise from the closer and optimised global integration of the process sub-units. This is one of the inherent commercial themes which motivated the research reported in this monograph.

To access the efficiency and feasibility of different large-scale system designs, the traditional tool has been the global steady-state analysis and energy balance. The process industries have many such tools encapsulated as proprietary design software. However, to obtain a vital and critical insight into global process operation a dynamic model and simulation is necessary. Over the last decade, the whole state of the art in system simulation has irrevocably changed. The Graphical User Interface (GUI) and icon based simulation approach is now standard with hardware platforms becoming more and more powerful. This immediately opens the way to some new and advanced large-scale *dynamic* simulation developments. For example, click-together blocks from standard or specialised libraries of process units are perfectly feasible now. For the industrial user, some questions arise: is it really as easy as the enthusiasts claim or do GUI based simulations have hidden development overhead costs? Should a company invest the resources to refit and refurbish existing software, or should it await the arrival of proprietary software libraries? How large a dynamic simulation can be managed and run in realistic time by the new GUI based software tools? The research reported in this monograph was an exercise in system simulation feasibility. A fairly large-scale process, a Combined Heat and Power plant was devised and the task was to produce a dynamic simulation. Overall the sum total of personnel input was about 40 person months of effort: this includes all the preliminary research, review and documentation work.

In the above discussion the *generic* features of the monograph have been emphasised. However, the applications area for the research are power generation plant, Combined Cycle and Combined Heat and Power. The objective of this phase of the overall research programme being the development of a process component software library where the blocks implement dynamic models. The outcome is a prototype simulation library of CC/CHP plants. The monograph covers all the development stages of the simulation from obtaining suitable models and assessing software tools to the construction of the simulation library and validation of process models blocks. Many of the questions above concerning the use of icon-based simulation tools were answered by the experience reported in this development. The question of global optimised power plant generation remains to be investigated in the second phase of this project.

A word about the nature of the collaboration is in order. The research is funded by the Engineering and Physical Sciences Research Council of the United Kingdom. Our companies John Brown Engineering Ltd (Clydebank), a Trafalgar House Engineering Division Company and Scottish Power plc are collaborating with the Industrial Control Centre of the University of Strathclyde in this research programme. Industry-university collaboration has some very useful benefits for both partners, for example, the exchange of technical knowledge, and project management to industrial standards or bench marks. This current collaboration continues a long and very fruitful association with the Centre at Strathclyde.

5th May, 1994

Colin Cloughley
Control Systems Manager
John Brown Engineering Ltd

Roger Farnham
Senior Engineer
Technology Division
ScottishPower plc

AUTHORS' PREFACE

The role of combined cycle and combined heat and electrical power generation systems has become more important over recent years due to both technological advances and the changing needs of the energy market. The greater flexibility, high efficiency and the relatively low capital cost of such systems has made them very attractive both for large and small power producing agencies.

Modern advanced simulation tools enable such systems to be rapidly analysed, aiding both the design process and the task of identifying improvements to existing systems. However, there is a need for the models to be available and for these to have the right structure and form to be useful in such simulations. Given a full simulation of such a power system it is possible to consider supervisory control, total system optimisation and also improved regulating loop controls. This monograph considers the first problem namely that of developing suitable models and selecting the simulation tools. There is some consideration of the control problem but this will be pursued in detail in a subsequent monograph. In fact the simulation which has been produced by the research enables different combined cycle and combined heat and power systems to be configured and analysed.

The results to be described were generated from a Engineering and Physical Sciences Research Council funded project with strong support from John Brown Engineering Ltd and ScottishPower plc. Our thanks are extended to Colin Cloughley and Roger Farnham, who have led the collaborative project for their respective companies.

The Monograph begins by considering the basic elements of CC and CHP systems. Models are described for each of the main components. In Chapter 2 the overall power generation control problem is considered dealing both with the lower level regulating loops and the high level supervisory control systems.

The important role of simulation tools is discussed in Chapter 3 and their value for this type of application is considered. The results of a brief review on different possible tools is presented.

A hypothetical flexible power system model described as the Skegton unit is considered in Chapter 4. This forms the basis of the major simulation tool which emerges from the research. The models considered in this section are for the major system components such as boiler, gas turbine and steam turbine. The chapter also includes a discussion on the control configurations used for the various subsystem components and the use of $MATRIX_x$ for simulating such systems.

Simulation results are then presented for individual subsystem components to demonstrate the validity of the results. The research reported in the Monograph will also be presented at a workshop to be given prior to the IEEE Conference on Control Applications to be mounted in Glasgow, Scotland in August, 1994.

Concluding remarks and a discussion on the way forward is presented in the final Chapter.

5th May, 1994

Andrzej Ordys, Andrew Pike
Michael Johnson, Reza Katebi,
Michael Grimble

ACKNOWLEDGEMENTS

The authors wish to acknowledge the financial support of the **Engineering and Physical Sciences Research Council** for the award of grant GR/H/44431. The support of the Electro-Mechanical Engineering Committee (Dr. Richard Liwicki, Secretary) and the continuing involvement of Douglas Warne (Committee Consultant) is kindly acknowledged.

The industrial impetus for the research is provided by ScottishPower plc (East Kilbride) and John Brown Engineering Ltd (Clydebank). The authors would like to thank Colin Cloughley (Control Systems Manager, JBE), Roger Farnham (SP) and David Reardon (SP) for their enthusiastic support, their managerial and technical input to the research.

The authors are very grateful for permission to publish the following:

- Fig. 1.11 – Permission kindly granted by Dr. John Waddington and POWERGEN, Ratcliffe-on-Soar, Nottingham, NG11 0EE, UK.
- Fig. 2.9 – Permission kindly given by Professor E. Welfonder, Department of Power Generation and Automatic Control, The University of Stuttgart, Stuttgart, Germany
- The material of Chapter 3 has previously appeared as the article *Simulation Tools for the 90's* by Andrew Pike and Michael Johnson, published in Measurement and Control, July, 1994. The kind permission of the Editor, Mr. T. Hughes (Institute of Measurement and Control) to reprint this material is kindly acknowledged.
- The authors would like to thank John McKendrick for the discussion on Distributed Control System Architectures and permission to use Fig. 2.1 and 2.2.

The authors wish to thank Mrs. Ann Frood and Mrs. Sheena Dinwoodie of the Industrial Control Centre for their excellent skills in typing the monograph and drawing figures.

Finally, any extended research task requires considerable dedication and this always leaves less time for family activities. The authors wish to thank their wives, children and friends for their patience and support during the writing of this monography.

The monograph makes reference to the following trademarks:

MATLAB® and SIMULINK® are registered trademarks of The Mathworks, Inc.

SPARC™ is a trademark of Sun Microsystems, Inc.

MATRIX$_x$® is a registered trademark of Integrated Systems, Inc.

SystemBuild™ and Xmath™ are trademarks of Integrated Systems, Inc.

X-Window™ is a trademark of M.I.T.
ACSL® is a registered trademark of MGA, Inc.
PROTOBLOCK™ is a trademark of Grumman Corporation.

EASY5$_x$® is a registered trademark of The Boeing Company.

SIMNON® is a registered trademark of Department of Automatic Control, Lund, Sweden.

VisSim™ is a trademark of Visual Solutions, Inc.

Windows® is a registered trademark of Microsoft Corp.

SIMAN ® and CINEMA® are registered trademarks of Systems Modeling Corp.

SIMPLE++™ is a trademark of AESOP GmbH.

ASPEN PLUS™, SPEEDUP™ and MAX™ are trademarks of Aspen Technology, Inc.

SIMFACTORY® is a registered trademark of CACI Products Company.

SIMPROCESS™ is a trademark of CACI Products Company.

WITNESS® is a registered trademark of AT & ISTEL.

Programe CC™ is a trademark of Peter M. Thompson and Systems Technology, Inc.

SPEEDTRONIC™ is a trademark of General Electric.

CONTENTS

EDITORIAL BOARD

CHAPTER 1

COMBINED CYCLE (CC) AND COMBINED HEAT AND POWER (CHP) SYSTEMS: AN INTRODUCTION

1.1 Introduction

Combined Cycle (CC) is a power plant system in which two types of turbines, namely a gas turbine and a steam turbine, are used to generate electricity. Moreover the turbines are combined in one cycle, so that the energy in the form of a heat flow or a gas flow is transferred from one of the turbines types to another. The most common type of Combined Cycle is where the exhaust gases from the gas turbine are used to provide the heat necessary to produce steam in a steam generator. The steam is then supplied to the steam turbine. However, as will be shown later, other connections between the gas turbine and the steam turbine are possible.

Combined Heat and Power (CHP) plant, sometimes called ***Cogeneration*** plant, are those which produce both electricity and heat. In this case electricity may be generated by a steam turbine or a gas turbine or, in smaller installations, by a diesel engine. Heat is extracted as steam or hot water and exported from the system. The ratio between electric output and heat output depends on the end requirements and this influences the selection of the configuration and sizing of components. The most common installation consists of a steam turbine cycle in which part of the steam is bled from the steam turbine to provide heat. However, other combinations are also possible. In particular a Combined Heat and Power scheme may also include a Combined Cycle installation.

The purpose of introducing Combined Cycle in power plants is to reduce losses of energy. Their main role lies in the utilisation of waste heat which may be found in exhaust gases from the gas turbine or at some other points of the process to produce additional electricity. In this way the overall efficiency of the plant - that is the efficiency of transformation of fuel energy into electric energy - increases reaching over 50%, when for traditional steam turbine plants it is approximately 40% and 35% for gas turbines plants [*Boissenin and Castanier, 1988*].

Similarly, in Combined Heat and Power the main objective is the utilisation of heat which otherwise would be wasted. This heat may be then supplied for district heating schemes or for industrial purposes. However, in some industrial and commercial applications Combined Heat and Power is used primarily for heat production. In that case the additional generation of electricity serves local needs and does not increase the overall efficiency of the system. Since the term Combined Heat and Power describes a more general type of installation, a wide variety of CHP schemes may be found. In Britain, the **Office of Electricity Regulation** has collected information for over 250 such schemes. Among them are large installations operating on steam turbines or gas turbines with an electric power output exceeding 100 MW, small packaged schemes with electric output less than 500 kW (typically consisting of gas fired internal combustion engines with heat recovery facilities) and also installations producing large amounts of heat in designated boilers with small additional production of electricity [*Marshall, 1993*].As examples of the latter, the installation owned by Slough Trading Estate Ltd. consists of :

Two circulating fluidized bed boilers,
One 23MW gas turbine with waste heat boiler,
Two 15MW steam turbines,
One 6MW steam turbine,
Three additional boilers.

The station produces 55MW electric power and 125MW heat from gas and oil fuels. Similarly, in Aberdeen, a power plant belonging to Wiggins Teape Fine Papers Ltd. includes a gas turbine, a boiler with supplementary firing and a back-pressure steam turbine. The plant uses gas and diesel oil producing 11.5MW electric power and 46MW heat.

Other examples of CC/CHP installations may be found in [*Jericha and Hoeller, 1991*], [*Yacobucci, 1991*], [*Boissenin and Castanier, 1988*], [*Wiggin, 1981*], [*Ahluvalia and Domenichini, 1990*]. According to [*Shields, 1989*] the largest Combined Cycle plant in the world is in Japan, where 14 gas turbines and 14 steam turbines (one gas turbine and one steam turbine driving one generator) generate 2000MW electric power firing natural gas.

This chapter is organised as follows :

- Sections 1.2, 1.3, 1.4, 1.5 give ***Basic configurations of thermal power plants,*** namely : ***Power Plants with Steam Turbines, Power Plants with Gas Turbines, Combined Heat and Power (CHP) Power Plants, Combined Cycle (CC) Power Plants***. This gives the background for the main types of possible CC and CHP schemes and corresponding flow (energy) diagrams. Differences and similarities are then discussed.
- In Section 1.6 the ***Main components of power system schemes*** are established and their structures are analysed. In particular, the most important input (control) and output (measurement) variables for each component are determined.
- Section 1.7 describes the ***Operational procedures currently employed***. This includes the operational regimes (as start-up, shut-down and normal operation), the main control loops for the system and for individual components, safety conditions and monitored variables.
- This material enables the construction of ***Block diagrams for the CC & CHP systems***, which are provided in Section 1.8.
- In Section 1.9, a ***Case study*** example is considered. For a fairly complicated CC/CHP configuration, the main control loops are established and certain control problems for both low level and supervisory control systems are addressed.

- Finally, the conclusions in Section 1.10 close the Chapter.

1.2 Power Plants with Steam Turbines

A typical example of a power train in a steam turbine power plant is presented in Fig 1.1. The scheme consists of :

- the boiler with furnace, drum, waterwalls (risers), superheater, reheater and economizer
- the steam turbine which may consist of high pressure section, intermediate pressure section and low pressure section
- the electrical generator
- the condenser
- the feedwater system (including deaerator)

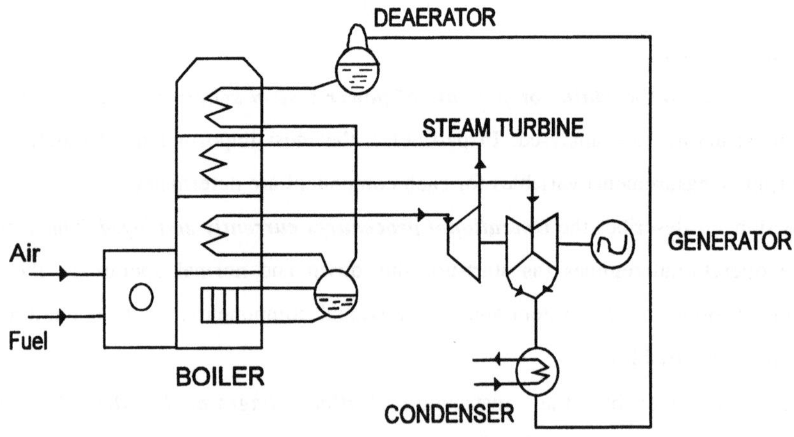

Fig. 1.1 Power plant with steam turbine, schematic diagram.

1.3 Power Plants with Gas Turbines

A typical example of a power train in a gas turbine power plant is presented in Fig 1.2. The scheme consists of :

- the air supply system to the compressor (inlet guide vanes)
- the fuel supply system to the combustor
- the turbine unit with the compressor, the combustor, and the turbine
- the electrical generator.

The high temperature exhaust gas is transported outside the system.

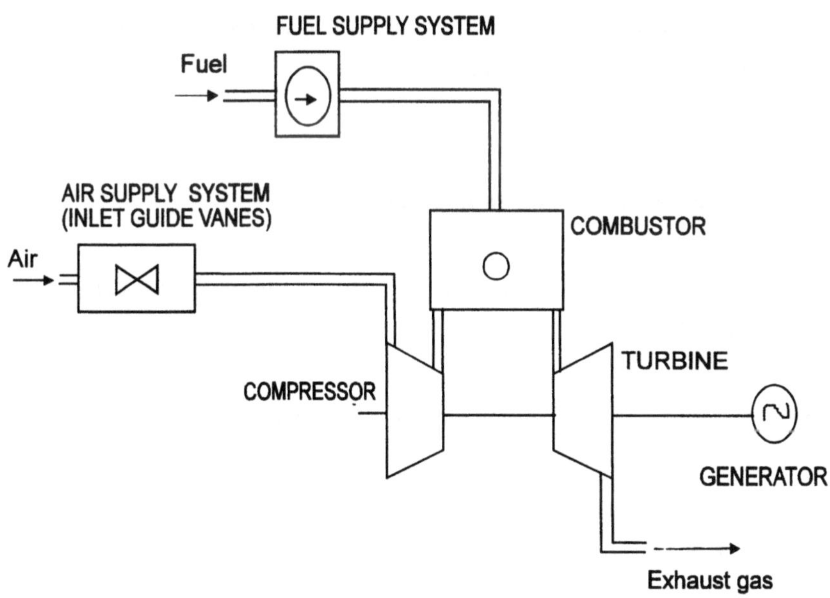

Fig. 1.2 Gas turbine power plant.

1.4 CHP Configurations

As was mentioned earlier, there exist a wide variety of CHP schemes. In this book the emphasis is on configurations which are primarily for electricity production. Consequently, the classification is based on the type of turbine and fuel used for electricity generation. Hence, three types of plants [*Basu and Cogger, 1986*] are considered : plants with steam turbines, plants with gas turbines and plants with diesel engines. For steam turbines schemes the fuel may be coal, natural gas or gas obtained from coal gasification. Liquid fuels are less popular due to their higher cost. For gas turbines, the fuel is mainly gas, however, liquid fuels may be used, especially during the start-up procedure when they are much safer. In addition, in CHP schemes with gas turbines an additional firing may be provided to the boiler and for this purpose any kind of fuel is suitable. Plants operating on liquid fuel are usually small local installations using modified diesel engines with added waste heat recuperation facilities. Also in this case an additional firing may be provided to the boiler.

1.4.1 CHP configuration with a steam turbine

A typical example of a CHP configuration with a steam turbine is shown in Fig. 1.3. The scheme consists of :

- Boiler, comprising the combustion chamber, the drum, the waterwalls (the risers), the superheater, the reheat (optional), and the economizer.
- Steam turbine, comprising high pressure section and two low pressure sections. The intermediate pressure section (optional) is not included in the diagram.
- Generator.
- Two valves, enabling the steam to bypass the high pressure section and the low pressure sections of the turbine and, in this way, provide heat for external users.
- Condenser.
- Deaerator.

In this scheme, heat is retrieved from the two steam circuits : heat is extracted at the reducing station from the steam by-passing the high pressure section of the turbine, and secondly from the the steam by-passing the low pressure section of the turbine.

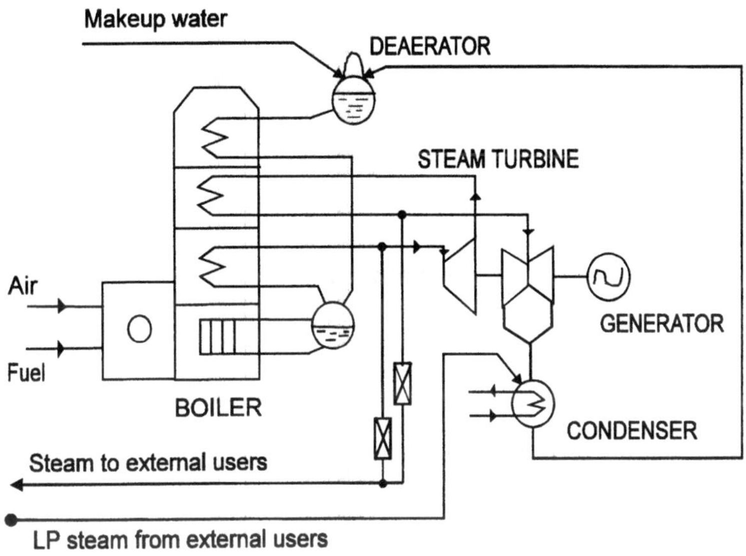

Fig. 1.3 CHP plant with a steam turbine.

1.4.2 CHP configuration with a gas turbine.

An example of CHP configuration involving a gas turbine is given in Fig. 1.4. The main components which may be identified on the diagram are as follows :

- Gas turbine comprising air compressor, combustor, turbine, connected to the generator.
- Turbine exhaust gases duct leading through to the boiler.
- Boiler, in this case called also Heat Recovery Steam Generator (HRSG).
- Steam cycle exporting heat from the CHP plant to external users.
- Feedwater system supplying water to the boiler.

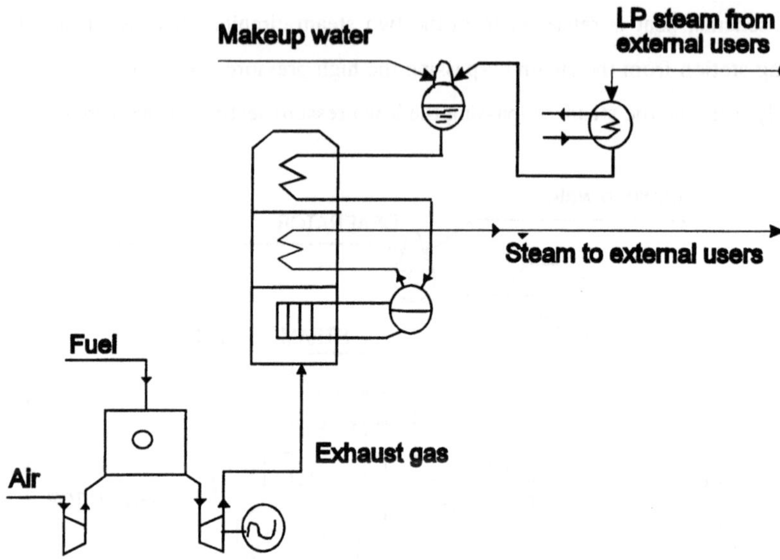

Fig. 1.4 CHP configuration with a gas turbine.

- Condenser (optional).

It is worth observing that such a configuration is suitable for supplying steam to a steam turbine, but, of course, the steam may be used also for other purposes. In that case, the condenser may not be required.

1.5 Combined Cycle Configurations

A classification of Combined Cycle power plants is given based on the publications [*Boissenin and Castanier, 1988*], [*Marecki, 1991*],. The three most popular types of CC power plants are :

- standard configuration (Fig. 1.5) : the exhaust gas from the gas turbine is used for steam production in the heat recovery steam generator. The HRSG extracts heat from exhaust gas to raise the steam parameters (temperature and pressure). The gas turbine plays the main role in electricity production.
- low excess air cycle (or hot windbox configuration) (Fig. 1.6) : in this case the exhaust gas from the gas turbine provides air to the boiler combustion chamber, the steam turbine is the main electricity producer (80-85%), [*Boissenin and Castanier, 1988*],
- CC with supplementary fired boiler (Fig. 1.7) : apart from recovering heat from the exhaust gas (as in standard configuration) an additional firing to the boiler is provided increasing the amount of produced steam. The electrical efficiency may be lower, when compared with the standard configuration, but there is an added flexibility in that the boiler may be supplied with different kind of fuel than the turbine. The proportion of electricity production which is divided between the gas part and the steam part depends on the level of additional firing.

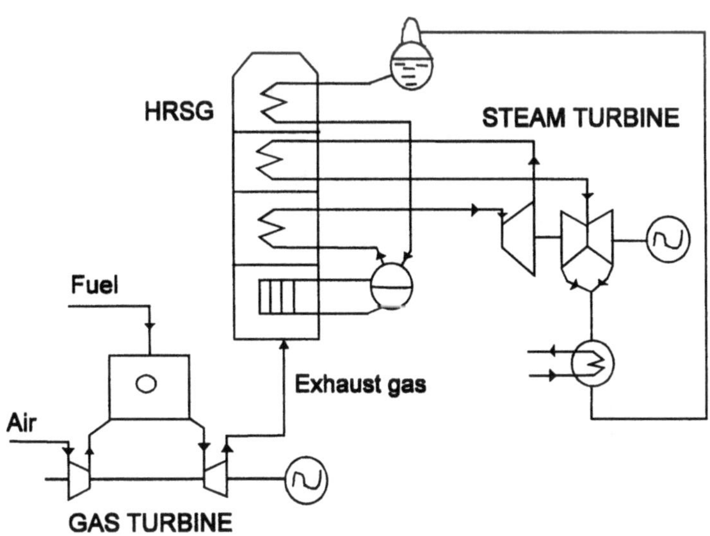

Fig. 1.5 Standard CC configuration.

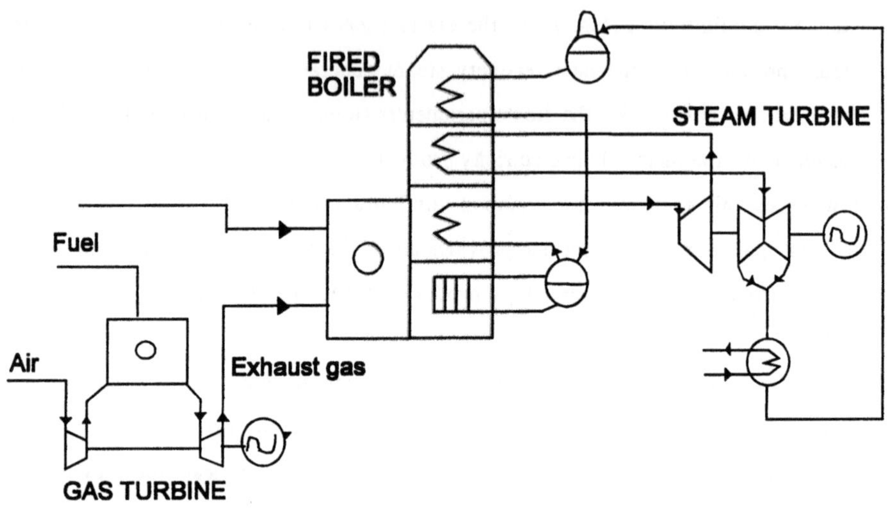

Fig. 1.6 Low excess air CC configuration.

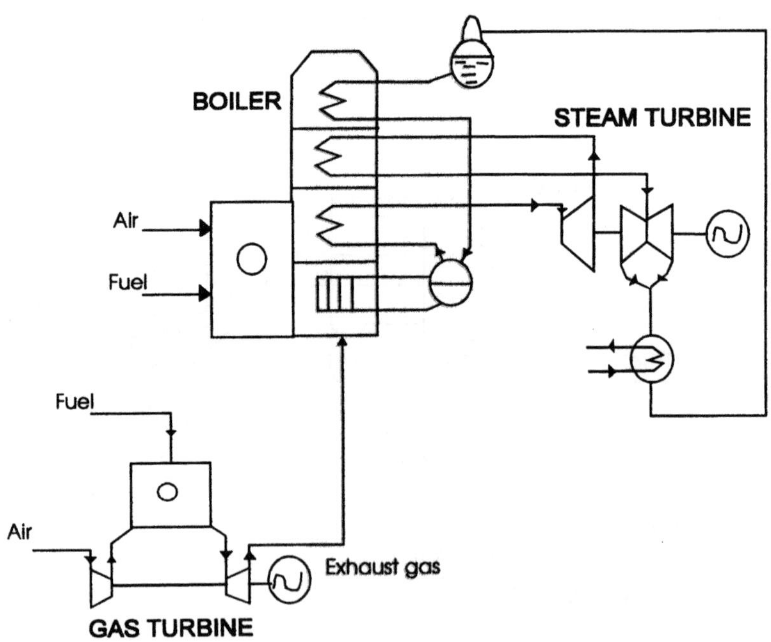

Fig. 1.7 CC configuration with supplementary fired boiler.

In addition to the above configurations a *reverse action* structure may also be considered (Fig 1.8). In it, the heat flow within the scheme is reversed in the sense that the heat is transported from the steam part (the boiler) to the gas part. The gas turbine is supplied either with gas or with clean air and the steam turbine plays the main role in electricity production.

Simplified diagrams of the CC power plants configurations have been presented in Figs 1.5 - standard configuration, 1.6 - low excess air cycle, 1.7 - supplementary firing, 1.8 - reverse action. The main components for these schemes are :

- Gas turbine (including combustion chamber and compressor) or diesel engine, in both cases with generator.
- Fuel and air supply to the gas turbine (or to the diesel engine).
- Steam turbine with generator.
- Condenser.

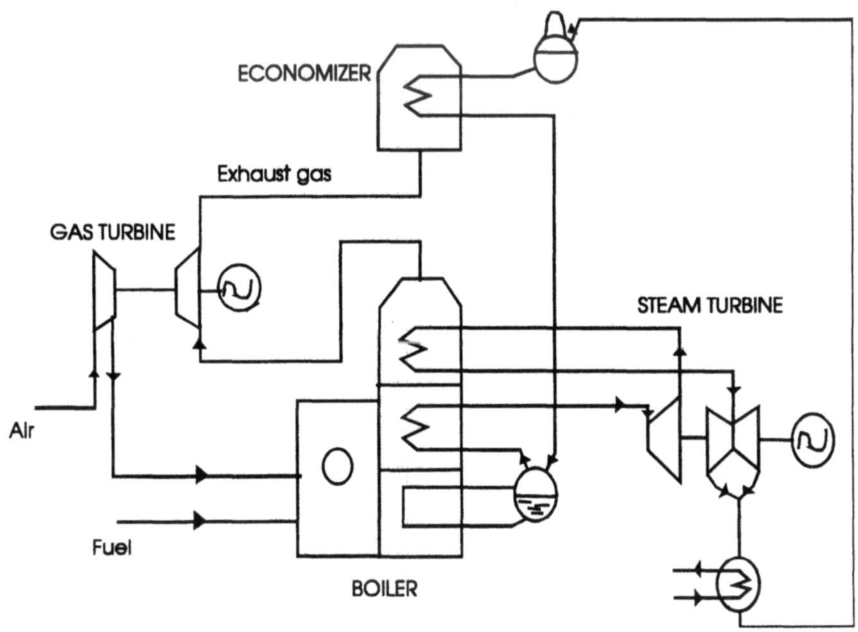

Fig. 1.8 Reverse action CC configuration.

- Boiler, unfired or fired (including a combustion chamber in the latter case).
- Fuel and air supply to the boiler (optional : only for fired boilers).
- Feedwater system of the steam turbine.
- Piping system together with valves, water and steam pumps, air compressors, water tanks.

1.6 The Main Components of Power Plants

1.6.1 Steam turbine

Basic types of steam turbines are as follows [*Bolland, 1991*], [*Marecki, 1991*] :

- condensing steam turbine : steam exits the turbine at almost vacuum conditions and enters a condenser,
- back pressure steam turbine : steam exits the turbine at a given pressure and may be used further (e.g. for heating),
- reduction steam turbine : a part of the steam is directed to bypass the turbine and may be used further,
- extraction steam turbine : a part of the steam is extracted from the turbine at a certain stage and may be used further,
- reheat steam turbine : a part of the steam is extracted from the turbine at a certain stage, is then reheated in the boiler and introduced back to the turbine,
- double pressure steam turbine : the turbine consists of two parts (two turbines working on a common shaft), namely high pressure turbine and low pressure turbine. The name refers to the steam pressure at the inlet to the turbine. In fact, the low pressure part usually consists of twin turbines, so that there will be three turbines rotating on a common shaft,
- triple pressure steam turbine : as above but, in addition, an intermediate pressure turbine appears.

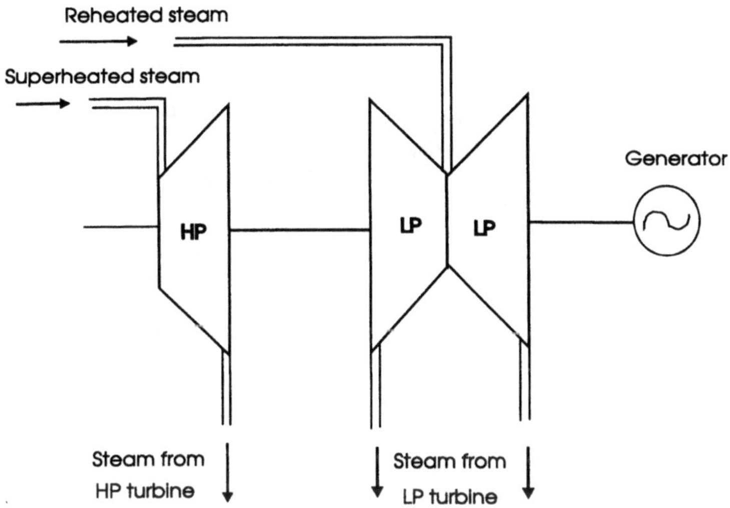

Fig. 1.9 Steam turbine - schematic diagram.

The schematic picture of the turbine which will be used in this book is given in Fig. 1.9. The diagram does not include the intermediate pressure section which is assumed to form part of the low pressure section. Thus, in terms of inputs and outputs, the diagram is generic enough, to cover all possible configurations. The models of different types of steam turbines may be found in IEEE Committee Reports : [-, *1972*], [-, *1991*]. The most important input and output variables are :

- parameters of steam (temperatures and pressures) at inlets and outlets of the turbine,
- electric parameters of the generator,
- certain safety measurements.

The number of inlets and outlets differs depending on the turbine type. In [*Domijan and Thompson, 1991*] an example of a steam turbine with 6 outlets is given. Typical values for steam parameters for modern heavy-duty steam turbines are as follows [*Shields, 1989*], [*Boissenin and Castanier, 1988*] :

high pressure turbine : input pressure: 50-160 bar, input temperature : 550°C,

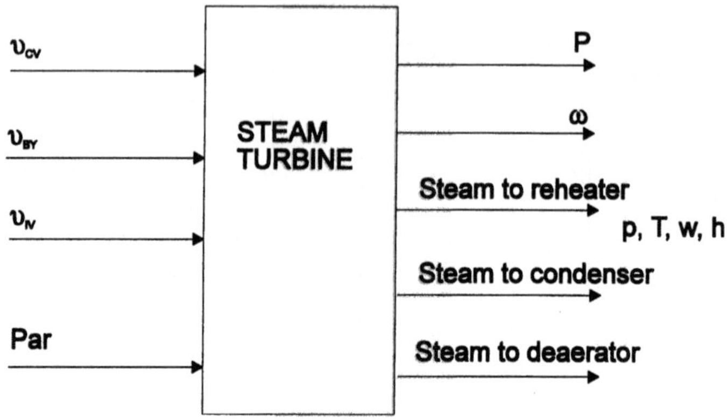

Fig. 1.10 Steam turbine - block diagram. Key : υ_{CV} - *control valve setting,* υ_{BY} - *bypass valve setting,* υ_{IV} - *intercept valve setting,* Par - *temperatures, pressures and enthalpies of inlet steam;* P - *mechanical power,* ω - *rotational speed,* p - *pressure,* T - *temperature,* w - *flow,* h - *enthalpy.*

intermediate pressure turbine : input pressure: 30-40 bar, input temperature : 550°C,

low pressure turbine : input pressure: 4-8 bar, input temperature : 250°C, output pressure: 95 mbar (when a condenser is used).

The electric parameters of the generator are output power (equivalent to load) and frequency (related to the speed of rotation). Safety measurements include vibrations of rotating parts of the turbine and temperature spread around the stator of the generator. Consequently, a steam turbine may be represented as the block diagram of Fig. 1.10.

1.6.2 Boiler

The boiler plays a crucial role in power plant equipment as a part which produces steam. In CC/CHP schemes two types of boilers may be found, they are :

- fired boilers,
- unfired boilers (heat recovery steam generator).

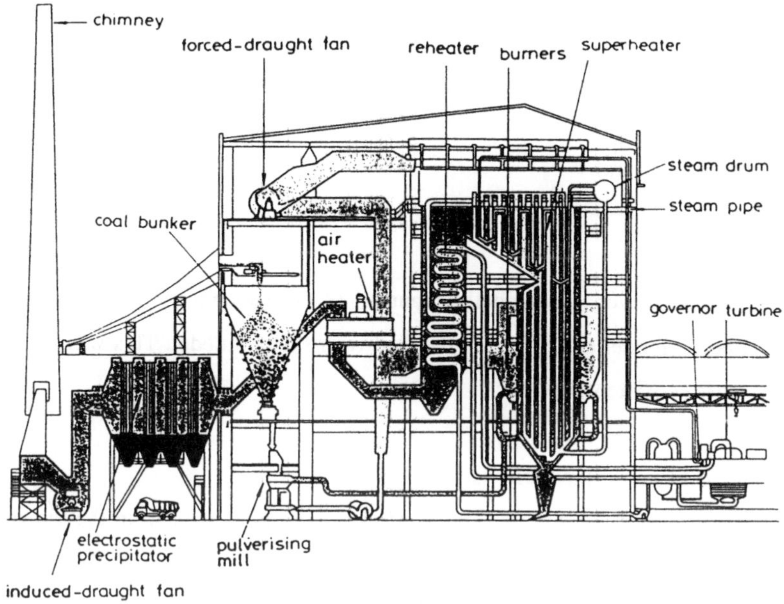

Fig. 1.11 A coal fired boiler [*Waddington and Maples, 1987*]

A steam generator with supplementary firing may be treated as a combination of the two above.

Fired boilers are used in power plants with steam turbines. In combined cycle, they appear when the participation of the steam part of the plant in electricity production is to be increased. The diagram of a coal fired boiler [*Waddington and Maples, 1987*] is shown in Fig. 1.11.

An unfired boiler, or heat recovery steam generator, is in fact a heat exchanger extracting heat from through-put gases and raising water to steam conditions. Its size is not as impressive as a fired boiler and it does not include a combustion chamber and a fuel supply system. However, the main characteristics of the steam raising components remain the same.

For the purpose of this book it is assumed that the boiler system may be divided into the following parts :

- Combustion chamber.
- Drum.
- Waterwalls (the risers).
- Superheater.
- Reheater.
- Economizer (however, for modelling purposes, the economizer will be removed from the boiler model and placed in the feedwater system model. This will make the models more consistent).

The schematic diagram of the boiler is presented in Fig. 1.12.

Mathematical models describing the features of boilers as seen from the viewpoint of control applications may be found in many publications [*Rovnak and Corlis, 1991*], [*Amin et al., 1984*], [*Chiacciarini et al., 1992*], [*Smith et al., 1984*], [*de Mello, 1991*], [*Patel, 1986*], [*Hogg and El-Rabaie, 1990*], [*Hogg and El-Rabaie, 1991*], [*Shields, 1989*], [*Waddington and Maples, 1987*]. The level of complexity of the models varies depending on a particular application area. The simplest models have two input and two output variables whereas the more complicated models have over ten inputs and outputs. Following [*Wildi, 1985*], the boiler may be divided into subsystems dealing with : water (feedwater, water drum), steam (drum, superheater), combustion and waste heat recovery, consequently the main variables are as follows :

water subsystem (the drum and the risers):

- inputs :
  - feedwater flow,
  - feedwater temperature,
  - superheater and reheater (attemporator) spray flows,
- outputs :
  - drum water level,

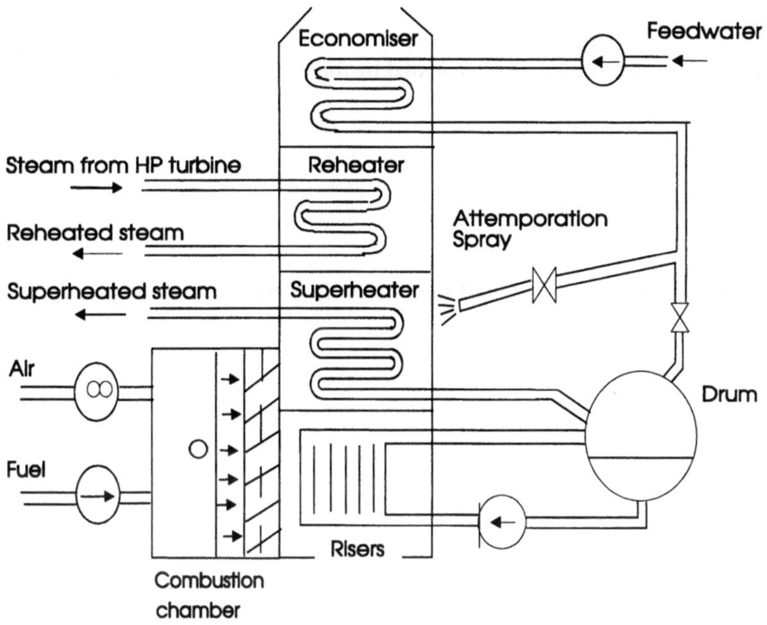

Fig. 1.12 Boiler - schematic diagram.

steam subsystem (the superheater and the reheater):

- inputs :
  - the inlet reheat steam parameters (temperature and pressure) - which represent an interconnection with the steam turbine,
- outputs :
  - superheat steam temperature,
  - superheat steam pressure,
  - reheat steam temperature,
  - reheat steam pressure,
  - boiler drum steam pressure,
  - boiler steam temperature,

combustion subsystem :

- inputs :

. fuel flow,

. air flow (or fuel to air ratio) - adjusted by force-draught fan speed,

. angle of furnace burner tilt,

- outputs :

. air pressure in furnace,

heat recovery subsystem (modelled together with the combustion system) :

- inputs :

. exhaust gas temperature,

. exhaust gas flow,

- outputs : none.

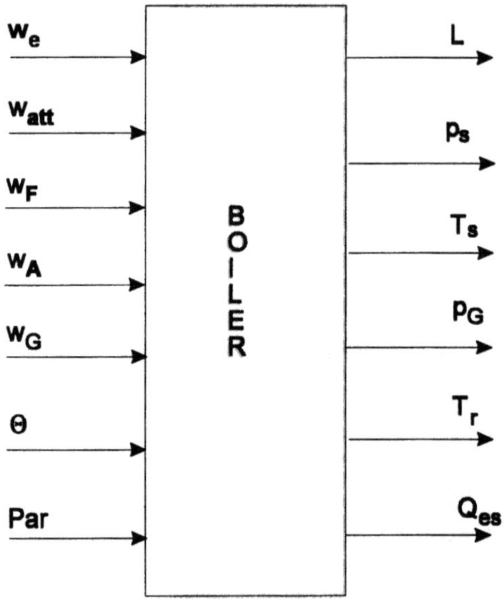

Fig. 1.13 Boiler - block diagram. Key : w_e *- feedwater flow*, w_{att} - *attemporator spray flow*, w_F - *fuel flow*, w_A - *air flow*, w_G - *exhaust gas flow from the gas turbine*, Θ - *tilt angle*, Par - *other input variables as follows : feedwater enthalpy, attemporator water enthalpy, economizer metal temperature, enthalpy of exhaust gas from the gas turbine, enthalpy of air, superheated steam flow to the turbine, reheated steam pressure, flow, temperature and enthalpy;* L - *drum water level*, p_s - *superheated steam pressure*, T_s *-superheated steam temperature*, p_G - *furnace air pressure*, T_r - *reheated steam temperature*, Q_{es} - *heat release to the economizer.*

Some of the variables listed above are internal and are only used in local control loops (e.g. drum water level, boiler steam pressure and temperature, air pressure in furnace). Others form connections between the boiler and the rest of the system. For the outlet steam, the parameters are obviously the same as the input parameters of the steam turbine connected to the boiler. The values for the remaining variables depend on the size of installation. For instance, the electric feedwater pump requires 1% of total electric power produced by the turbine to supply water to the boiler. A block diagram description of the boiler is given in Fig. 1.13.

1.6.3 Gas turbine.

The main components of gas turbine are : compressor (17 stages) with air inlet, combustor with fuel supply system (usually consists of two individual subsystems - gas fuel system and oil fuel system), 3-stage turbine, exhaust gas duct, cooling system (supplied with lubricant oil or water). Contemporary gas turbines produce electric power in range up to 200 MW, and the outlet gas temperature is around 550°C when the combustion gas temperature reaches 1200°C. Fluid flows depend on the power produced (e.g. for the gas turbine GE Frame 9 the fuel flow is around 9kg/s, the air/fuel ratio is 670% and the water injected to the combustor is around 3.5kg/s at full load). Models and parameters of gas turbines may be found in [*Shields, 1989*], [*Rowen, 1983*], [*Rowen, 1988*], [*Hussain and Seifi, 1992*], [*Yacobucci, 1991*]. The classification of different types of gas turbines includes [*Cohen et al., 1987*]:

- single shaft gas turbine,
- double shaft gas turbine,
- twin spool gas turbine.

These may be operated in an open or closed thermodynamic cycle. The compressor and the turbine may have a single stage or be segregated into multiple stages corresponding to different operating pressures. Complex plant arrangements include intercooling, heat-exchange and reheat.

The schematic diagram of the gas turbine (as used in this book) is presented in Fig. 1.14.

The main input and output variables for the gas turbine are as follows [*Ahluvalia and Domenichini, 1990*], [*Yacobucci, 1991*] :

- inputs :
  - fuel (or fuels) flow,
  - air flow (or air/fuel ratio),
  - air temperature,
  - water injection flow,
  - parameters of the cooling system,
- outputs :
  - electric power produced (or load),
  - frequency (speed of rotation),
  - exhaust gas flow,
  - exhaust gas temperature,

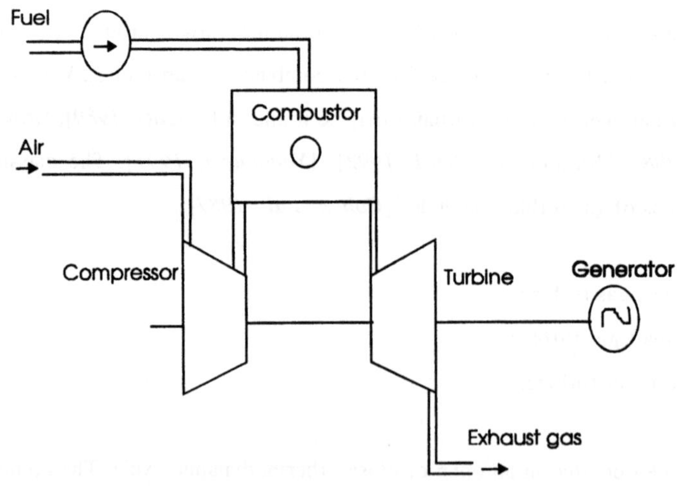

Fig. 1.14 Gas turbine - schematic diagram.

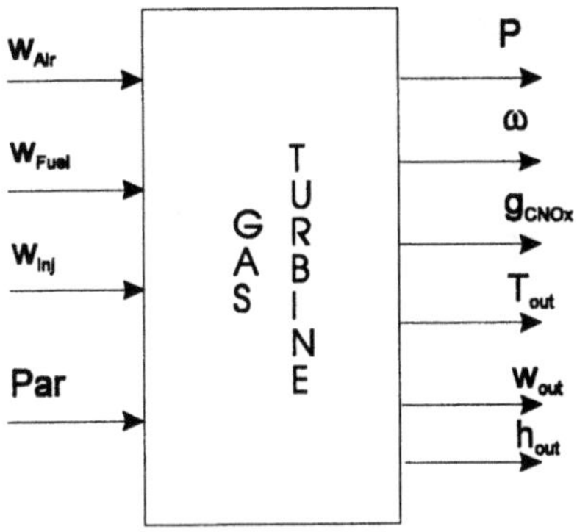

Fig 1.15 Gas turbine - block diagram. Key : w_{Air} - *air flow to the turbine*, w_{Fuel} - *fuel flow to the turbine*, w_{inj} - *steam (water) injection flow*, Par - *enthalpy, density, pressure and temperature of inlet air, temperature of injection liquid*; P - *mechanical power*, ω- *rotational speed*, g_{CNOx} - *NO_x content in exhaust gases*, T_{out} - *exhaust gas temperature*, w_{out} - *exhaust gas flow*, h_{out} - *exhaust gas enthalpy.*

- exhaust gas O_2 content,
- safety measurements for the generator such as vibrations of the rotor and temperature spread around the stator.

A block diagram of gas turbine is given in Fig. 1.15.

1.6.4 Condenser

The condenser is situated at the steam outlet of a condensing type of steam turbine. A condenser causes the steam to condense by letting it flow over cooling pipes. Cold water from an outside source flows through the pipes thus carrying away the heat. A condensate pump removes the condensed water and drives it to boiler feedwater pump. About one-half of the energy produced in the boiler is removed from the steam when it exhausts into the condenser. The temperature of the cooling water increases typically by 5°C to 10°C. The condensed steam (condensate) usually has a temperature between 27°C and 33°C and the absolute pressure about 50mbar [*Wildi, 1985*]. Descriptions of condensers may be found in [*Shields, 1989*], [*Bolland, 1991*]. In [*Kapadia and*

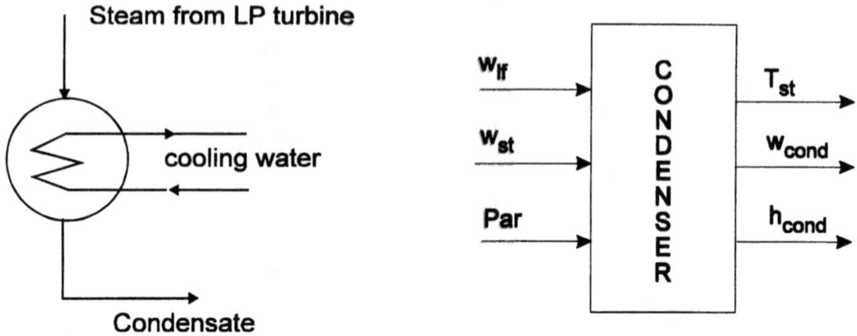

Fig. 1.16 Condenser, schematic diagram and block diagram. Key : w_{lf} - *cooling liquid flow*, w_{st} - *inlet steam flow*, Par - *inlet steam enthalpy, cooling liquid temperature*; T_{st} - *outlet condensate temperature*, w_{cond} - *outlet condensate flow*, h_{cond} - *outlet condensate enthalpy.*

Wolgemuth, 1984] more detailed considerations, including a model with many internal states, are given. However, for the purpose of this research, the following input and output variables will be considered :

- inputs :
  - steam pressure,
  - steam temperature,
  - steam mass flow,
  - cooling water flow,
  - cooling water inlet temperature,
- outputs :
  - condensate temperature,
  - condensate flow,
  - cooling water output temperature.

A schematic diagram and block diagram for a condenser as used in this book are depicted in Fig. 1.16.

1.6.5 Feedwater system

A schematic diagram of the feedwater system is depicted in Fig. 1.17. The feedwater system consist of the following parts : deaerator, water pump (the main feedwater pump), economizer, feedwater valve, attemporation valve. The main inputs and outputs are as follows :

- inputs :
  - makeup water flow,
  - makeup water temperature (or enthalpy),
  - condensate flow,
  - condensate temperature (or enthalpy),
  - injected steam flow,
  - injected steam temperature (or enthalpy),
- outputs :
  - deaerator water level,
  - deaerator steam pressure,
  - feedwater flow,

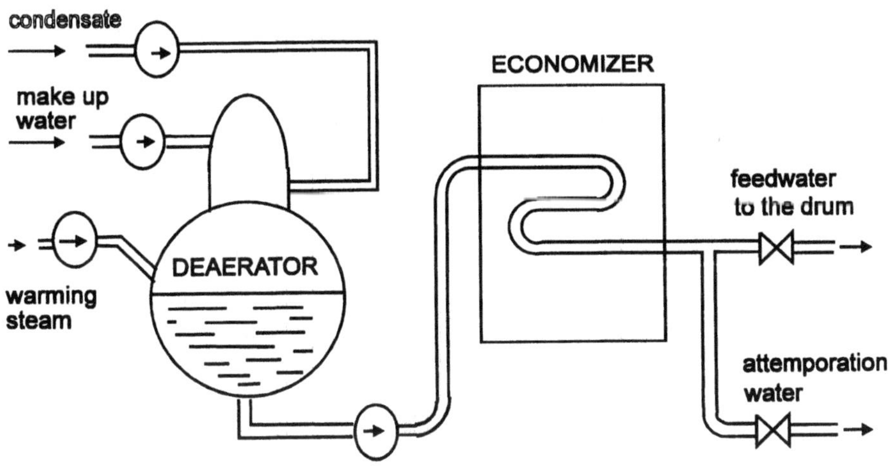

Fig. 1.17 Feedwater system - schematic diagram.

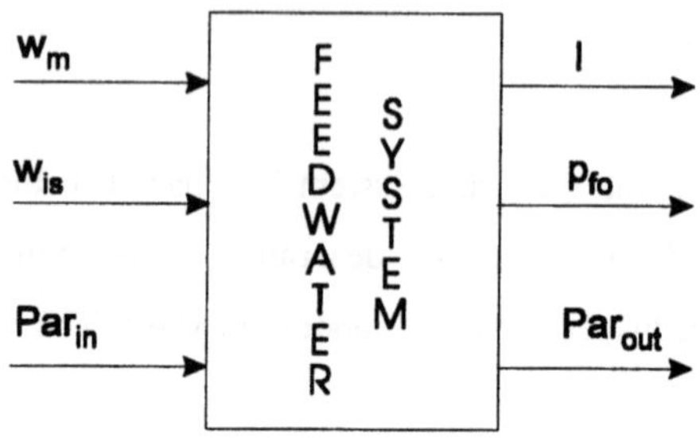

Fig. 1.18 Feedwater system - block diagram. Key : w_m - *makeup water flow*, w_{is} - *injected steam flow*, Par_{in} - *other input variables like enthalpies, flows*; l - *deaerator water level*, p_{fo} - *deaerator steam pressure*, Par_{out} - *other output variables like flows, temperatures, enthalpies.*

- feedwater temperature (or enthalpy),
- attemporation water flow,
- attemporation water temperature (or enthalpy).

The block diagram of the feedwater system is presented in Fig. 1.18.

1.6.6 Miscellaneous components

Apart from the components described above, a power plant includes hundreds of other elements. Among them, the most important from the point of view of control strategies are : valves, pumps, air compressors and piping systems. Some of them will be included in models of bigger components (e.g. feedwater system) whereas for others it may be more convenient to introduce separate models providing additional flexibility in the control strategy. In this case, a simple description will be used although some nonlinearities, for example saturation, must be taken into account. Pipes and valves are modelled in [*Stout, 1984*] and [*Orchard and Robinson, 1981*] considers a steam network.

1.7 Operational Procedures

1.7.1 Operational regimes

In general, three stages of operation of power plants or its components are distinguished. They are (see e.g. [*Rowen, 1988*]) :

- start-up regime,
- shutdown regime,
- normal (steady state) regime.

In normal operation regime a power plant may work independently of the rest of the national grid - so called island operation mode, or may be connected to the network (national grid) - so called parallel operation mode [*Ahluvalia and Domenichini, 1990*]. In the parallel operation mode, the requirements for the power generating units are specified by system operators (load dispatchers) and the energy management centres. This again puts a power generating unit into one of the following conditions [*Conradie and Kürten, 1992*] :

- constant power generation,
- generation cycling (scheduled operation at and between given levels of power for specified periods),
- generation for system frequency support in the short term,
- generation in a spinning reserve mode with frequency error compensation.

It is important to note that control loop assignment and set point specifications may differ for these different conditions.

Start-up and shut-down procedures are the most challenging problems for control applications. They require the sequence of operations to be successfully performed leading a turbine and associated power plant components through a sequence of safe states. At the same time many variables must be monitored and checked to ensure safety of operation. Moreover, minimum time and minimum energy losses during the procedures would be desirable. The start-up procedure for a modern heavy-duty steam turbine lasts for around 6 hours. It consists of the following stages [*Choudhari, 1986*] :

1. Warming up of main steam pipe line.
2. Warming up of turbine parts.
3. Turbine run-up.
4. Synchronisation.
5. Loading.

For a gas turbine, a start-up sequence is described in [*Rowen, 1988*]. The stages are as follows :

1. Pumping lubricant oil to produce an oil layer under the rotor of the turbine and, in this way, to decrease static friction.
2. Using a starter motor to begin the rotation of the turbine.
3. Supply oil fuel to the combustion chamber and establish a stable flame.
4. After achieving required rotation speed and exhaust gas temperature - synchronisation to the network.
5. Gradually changing from oil fuel to gas fuel.
6. Increasing load to the required value.

Additional problems are involved when starting-up or shutting down certain components within a bigger power station. This is especially difficult for Combined Cycle plants because, in this case, interconnections between components occur not only through the electric network but also on the "heat exchange" side, i.e. flows of gases, and flows of steam. In [*Nakanomori and Kano, 1992*] examples of start-up procedures for a combined cycle plant consisting of two gas turbines, two heat recovery steam generators and one steam turbine are described :

First, preparation process start for steam generation, such as drum water filling and vacuum raising. Second, gas turbine start-up sub-sequence and the first heat recovery steam generator simultaneous start-up sub-sequence follow. The heat recovery steam generator is pressurised with heat input increase by gas turbine load program control. Finally, the steam turbine start-up sub-sequence runs up, synchronises and loads the steam turbine while the steam turbine stress is kept within the limits by controlling the steam temperature.

Differences for other mode transitions are as follows. For the heat recovery steam generator start-up with gas turbine in operation, heat input is increased by the damper set point program control. For the second heat recovery steam generator start-up, the connection to the operating steam turbine is additionally carried out with the steam temperature matching control. For most of the shutdown mode transitions, the sequence is advanced with the corresponding start-up mode transition in the reverse order. For the gas turbine/heat recovery steam generator shutdown, heat input to the heat recovery steam generator from the gas turbine is cut by diverting gas turbine exhaust gas flow to the atmosphere before initiating gas turbine shutdown sequence.

In [*Kano, 1992*] some problems of the control of a boiler in a low excess air combined cycle occurring during start-up and shutdown of a gas turbine connected to it are described. Requirements for a so-called changeover control system are as follows :

- quick start-up of the forced draft fan and fast damper changeover to enable continuous firing when the gas turbine is shut-down,
- smooth damper changeover control without losing air from the forced draft fan to the bypass exhaust stack,
- reverse operation from the forced draft fan to the gas turbine exhaust when the gas turbine is ready again,
- fuel cutback corresponding to the decreasing combustion air during the changeover to avoid unburned fuel explosion,
- steam generation back-up by other boilers.

A more sophisticated approach to the operational regimes, so called boundary operation regimes, is developed in [*Welfonder, 1992*]. The boundary operation regimes are described in a n-dimensional space as an area where given system variables approach one or more safety constraints. When this happens control algorithms have to react to avoid constraints violation and assure smooth operation. For steam plant, the variables chosen for this purpose are :

- power parameters : frequency, voltage, power flow,
- generator parameters (e.g. rotation speed),
- steam parameters : pressure, temperature.

1.7.2 Main control loops

There are two broad categories of control systems in power plants : sequence control and modulating control.

1.7.2.1 Sequence control

Sequence control is associated only with the start-up or shutdown of items of plant. A sequence ensures, for instance, that a large fan or fuel supply system cannot be started until a supply of lubricating oil has been provided for the bearings. With large items of plant, the associated sequences can be complex and it is essential that they should operate reliably if the overall plant performance is to be adequate to meet the system demand. The main problems to be overcome are the unreliability of signals and the diagnosis of failures [*Waddington and Maples, 1987*].

1.7.2.2 Modulating control, main control loop

Modulating control aims to make adjustments to plant which is already operating in order to maintain parameters at desired values. The main control task in power plant is to control the balance between generator and load. This is realised by measuring and correcting the actual frequency [*Wildi, 1985*], [*Cegrell, 1986*]. If the frequency increases it means that the power produced is above the demand. Similarly, a frequency drop indicates that the power production is not sufficient. The main control loop in power plants is called the governor control. Consequently, frequency stabilisation is the main control task. It is possible to change the load dispatch using the governor control loop. For this, the set point, i.e. the desired frequency has to be set slightly different than in the grid. It involves changes in load allocated to the unit.

Governor control can be realised by different means. Available solutions for gas turbines are described in [*Honderd et al., 1992*], [*Yacobucci, 1991*], [*Rowen, 1988*], [*Dolbec, 1984*]. In [*Honderd et al., 1992*] load and frequency are adjusted by changes in fuel flow. The fuel flow is also used to control the inlet gas temperature. In [*Yacobucci, 1991*] the fuel flow regulates load and frequency and at the same time, the exhaust gas temperature. Moreover, the temperature control system can override the frequency or load control. The turbine fuel system consist of two subsystems : gas fuel system and oil fuel system. Most gas turbines also use inlet guide vanes to reduce load down to around 80% and maintain high efficiency [*Brown, 1993*].

In steam-turbine power plants, governor control is usually considered for the group consisting of steam turbine component and boiler component. This leads to the possible configurations as follows [*Conradie and Kürten, 1992*] :

- turbine-following mode (constant boiler pressure) : the component load command adjusts the fuel supply to the boiler and the turbine admission valves control the boiler pressure,
- boiler following mode (sliding boiler pressure) : the component load command adjusts the turbine admission valves by means of the speed or power controller and the steam pressure is controlled by the fuel supply to the boiler,
- unit coordinated mode : the component load command is applied simultaneously to both, the turbine admission valves and the fuel supply system, the steam pressure is trimmed by the turbine admission valves and/or by the fuel supply to the boiler,
- frequency compensation mode : requires the provision of a system frequency compensation device to modify the power demand signal by a bias proportional to frequency deviation from a target value set by the system operator.

In [*Ahluvalia and Domenichini, 1990*] two modes of operation related to the main control loop in combined cycle in parallel operation with the national grid are considered. They are :

- power control,
- droop control on speed.

In the power control case, the measured variable is the total power produced by the cycle, in the droop control case, the measured variable is the machine speed, which is set by the national grid.

1.7.2.3. Modulating control, other control loops

Apart from governor control, there are many control loops for the individual components of power plants. Only the most important are listed below.

Gas turbine controls [*Rowen, 1988*], [*Yacobucci, 1991*] :

- control of fuel flow to achieve a desired load,
- control of fuel/air ratio, to provide a correct output gas temperature,
- control of injected water flow to avoid NOx emission,
- control of lubricant oil flow and temperature,
- control of inlet guide vanes.

Steam turbine controls [*Ahluvalia and Domenichini, 1990*], [*Welfonder, 1992*], [*Kure-Jensen and Hanisch, 1991*] :

- high pressure steam valve control to adjust speed and power,
- other steam flow valves control to adjust pressures in the turbine,

Boiler controls [*Waddington and Maples, 1987*], [*Amin et al., 1984*], [*Keeton, 1984*], [*Yang and Hogg, 1992*], [*Ahluvalia and Domenichini, 1990*], [*Dwarakanath et al., 1982*], [*Welfonder, 1992*], [*Mann and Lausterer, 1992*], [*Hogg and El-Rabaie, 1991*] :

- control of fuel supply to the boiler to adjust power and speed of steam turbine,
- control of fuel supply to the boiler to adjust the boiler drum steam pressure,
- regulating a feedwater pump speed to maintain the drum water level,
- control of induced draught fan speed to adjust the air pressure in furnace,
- control of superheat furnace burner tilt to adjust the superheat steam pressure,
- control of superheat spray water flow to adjust the superheat steam temperature,
- control of reheat furnace burner tilt to adjust the reheat steam pressure,
- control of reheat spray water flow to adjust the reheat steam temperature,
- changing air damper position to control air flow through the boiler (and exhaust gas temperature).

Condenser controls :

- control of cooling liquid flow to maintain parameters of condensate.

Feedwater system controls :

- control of water level in the condensed water tank (deaerator),
- control of steam injection flow to maintain the deaerator pressure.
- adjustment of valves to provide desired outlet water flow.

Other controls :

- control of different valves positions to adjust steam flows distribution,
- adjustment of gas damper to change the exhaust gas flow through the boiler.

In total, modulating control takes care of hundreds of variables, only a few of them are mentioned above. Moreover, actions which are described above as single loops may consist of many sub-systems, for instance a valve position is actuated by an electric motor and adjusting three currents (for three coils) is required. Reliability of the system is essential. This means, in particular :

- duplication (or even triplication) of all control loops to prevent malfunctioning in case of a controller failure,

- availability of tools for easy detection of failures.

Consequently, the idea of integrated control systems for each component, which is believed to give more compact and more robust approach, is widely implemented in practice [*Alvarez-Calderon, 1984*], [*Kure-Jensen and Hanisch, 1991*]. A power plant is divided into several hierarchical levels. On each level certain components with separate control algorithms but with possible interconnections between components are distinguished. For example, for a simple Combined Cycle plant, components are : steam turbine with associated control loops, boiler with its control loops, gas turbine with a control system (e.g. Speedtronik Mark IV), auxiliary controls. A power group master control system includes all these components and provides a supervisory control (set points) for them when, at the same time being controlled by a network managing control system. It is also possible to go down in the hierarchy dividing for example a boiler control component into several control sub-components corresponding to control loops as listed above.

1.7.3. Safety conditions

As mentioned earlier, the protection system plays crucial role in operation of both gas cycle and steam cycle in power plant.

For gas turbines the following conditions, if detected, indicate the development of conditions approaching safety constraints [*Rowen, 1988*], [*Yacobucci, 1991*] :

- excessive turbine speed,
- excessive turbine exhaust gas temperature,
- excessive turbine vibration,
- a loss of flame,
- excessive lubricant oil temperature,
- low hydraulic pressure,
- low pneumatic pressure,

- low lubricant oil pressure,
- excessive generator temperature,
- compressor surge.

For steam turbines, safety conditions are as follows [*Kure-Jensen and Hanisch, 1991*], [*Amendola et al., 1986*] :

- excessive turbine speed,
- excessive turbine vibration,
- loss of lubricant oil pressure,
- high lubricant oil temperature,
- excessive bearing vibrations,
- valves vibrations and failures.

For boilers, safety requirements result in monitoring procedures to protect against the following conditions occurring [*Amendola et al., 1986*], [*Kure-Jensen and Hanisch, 1991*] :

- excessive steam pressure in the boiler drum,
- excessive superheat or reheat steam temperature,
- low drum water level,
- chemical anomalies in the water-steam cycle,
- a loss of flame,
- incorrect air pressure in the combustion chamber,
- thermal stresses.

Additionally, measurements of :

- internal deposit in the water-wall tubes of the boiler,
- material ageing

are performed.

Also for other power plant components some safety measurements are realised, these include :

- feed pump and air fan vibrations,
- efficiency of thermal exchange in the condenser and in the heat exchanger,
- detection of leakage.

1.8 Block Diagrams for the CC & CHP Systems

The diagrams of power systems, including CC and CHP configurations, presented in sections 1.2, 1.3, 1.4, 1.5 are flow or energy diagrams. For control purposes it would be useful to construct diagrams consisting of blocks as depicted in section 1.6. Such diagrams are discussed and devised next.

1.8.1 CHP installation with steam turbine

In Fig. 1.19 a block diagram is presented for a CHP installation with steam turbine. The diagram consists of four blocks : steam turbine, boiler, condenser and feedwater system. Interconnections between the blocks are as follows :

- output from the boiler (a vector of steam parameters) is connected to :
  the input to the steam turbine,
  the input to the feedwater system,
- output from the steam turbine (a vector of steam parameters) is connected to :
  the input to the condenser,
  the input to the boiler (reheat steam parameters),
- output from the condenser is connected to :
  the input to the feedwater system,

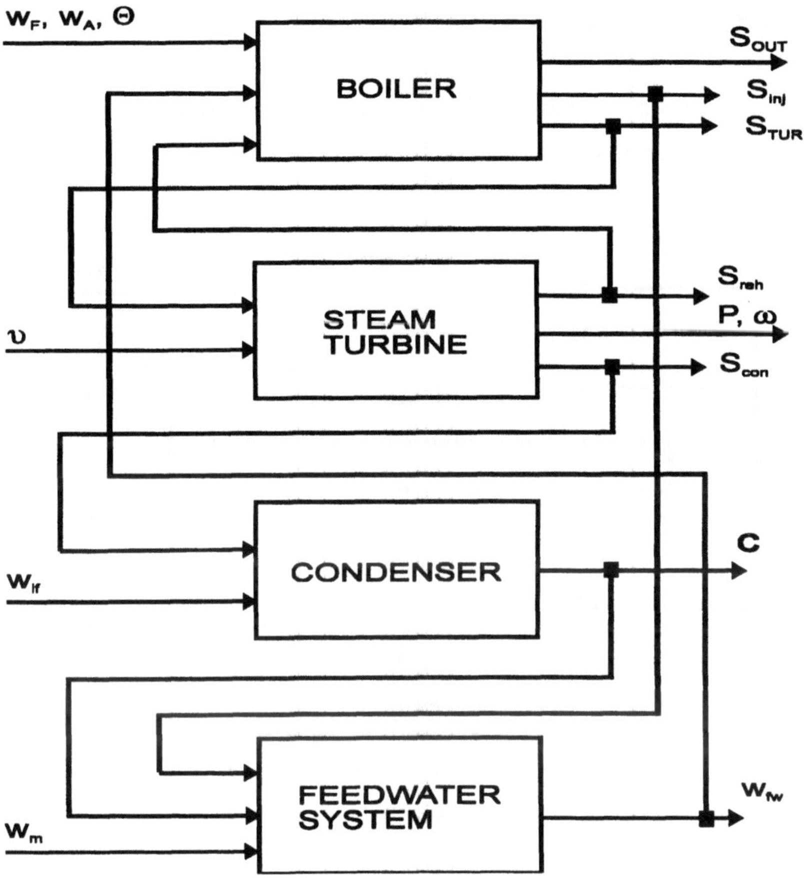

Fig. 1.19 CHP installation with steam turbine - block diagram.

- output from the feedwater system is connected to :

    the input to the boiler.

The heat for external users is extracted from the output from the boiler or (and) from the output from the steam turbine.

1.8.2 CHP installation with gas turbine (or diesel engine)

Fig. 1.20 shows CHP with gas turbine or diesel engine. The boiler (heat recovery steam generator) may have additional firing provided. The main interconnections are :

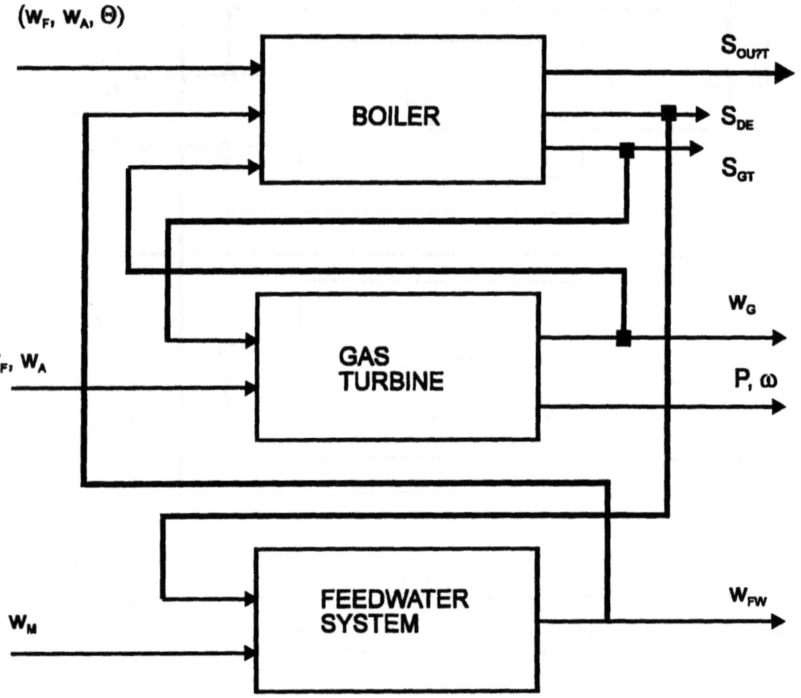

Fig. 1.20 CHP installation with gas turbine (or diesel engine) - block diagram.

- output from the gas turbine (a vector of exhaust gas parameters) is connected to :
  the boiler,
- output from the boiler (a vector of steam parameters) is connected to :
  the feedwater system (steam injection to the deaerator),
  the gas turbine (steam injection to the combustion chamber)
- output from the feedwater system (a vector of water parameters) is connected to :
  the boiler.

In this scheme, the heat for external users is extracted only from the boiler. After extracting the heat, the water returns to the feedwater system.

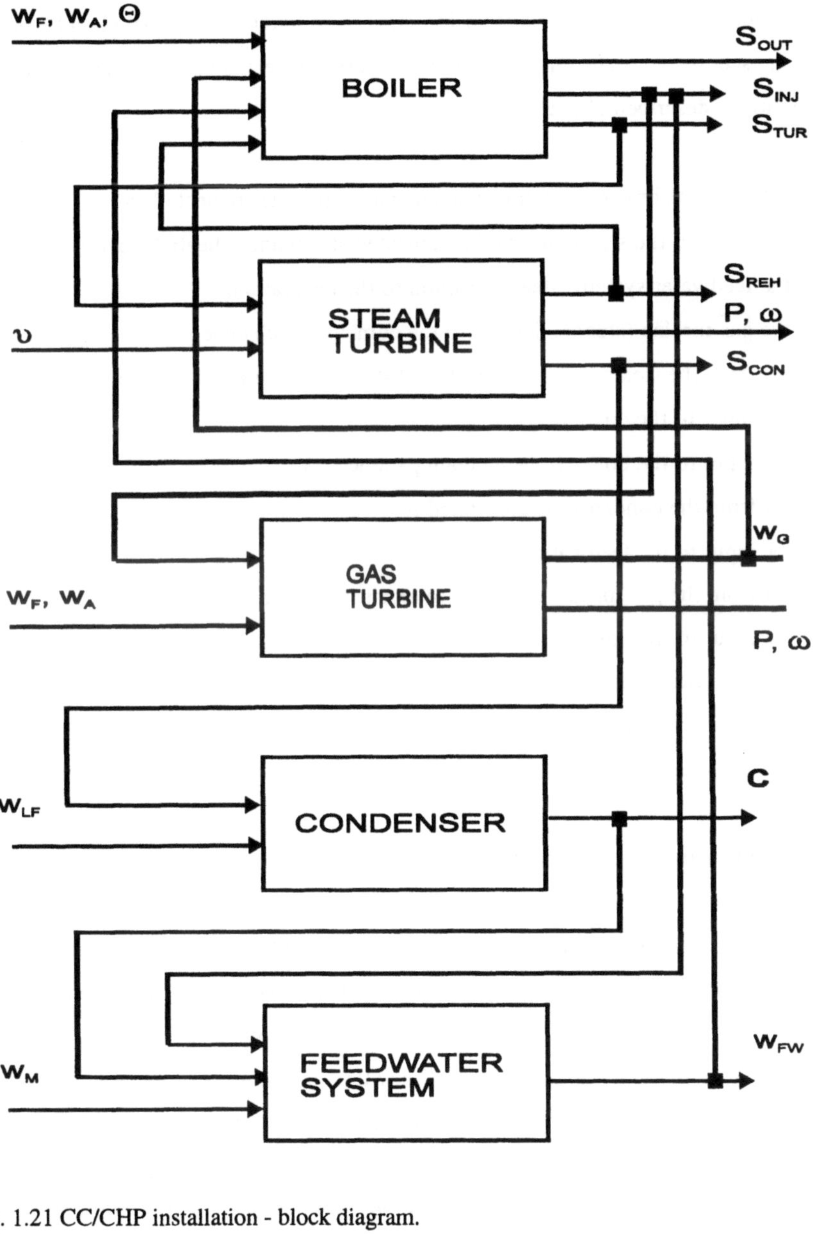

Fig. 1.21 CC/CHP installation - block diagram.

1.8.3 CHP with CC process

A Combined Heat and Power including Combined Cycle (both gas turbine and steam turbine) block diagram is presented in Fig. 1.21. The diagram may include all elements listed in section 3. The interconnections between elements are similar as for configurations described above :

- output from the boiler (a vector of steam parameters) is connected to :
  the input to the steam turbine (superheated steam and reheated steam),
  the feedwater system (steam injection to the deaerator),
  the gas turbine (steam injection to the combustion chamber)
- output from the steam turbine (a vector of steam parameters) is connected to :
  the input to the condenser,
  the input to the boiler (reheat steam parameters),
- output from the condenser is connected to :
  the input to the feedwater system.
- output from the gas turbine (a vector of exhaust gas parameters) is connected to :
  the input to the boiler,
- output from the feedwater system is connected to :
  the input to the boiler.

1.8.4 A component based block diagram

In Fig. 1.22 a more general approach is presented. A Combined Cycle Combined Heat and Power plant is divided into separate components like boiler, steam turbine, gas turbine, condenser, feedwater system, or any of miscellaneous elements listed in section 1.3. However, more than one component of any type may appear on the diagram. Thus, the total number of components depends on plant complexity. Here it is denoted by **n**. Interconnections between components result from CC/CHP technology as presented in section 1.8.3, for instance, the gas turbine exhaust gas parameters are connected to the input to the boiler. However, if more than one component of one type appears on the diagram, e.g. two boilers, then interconnections will become more complex.

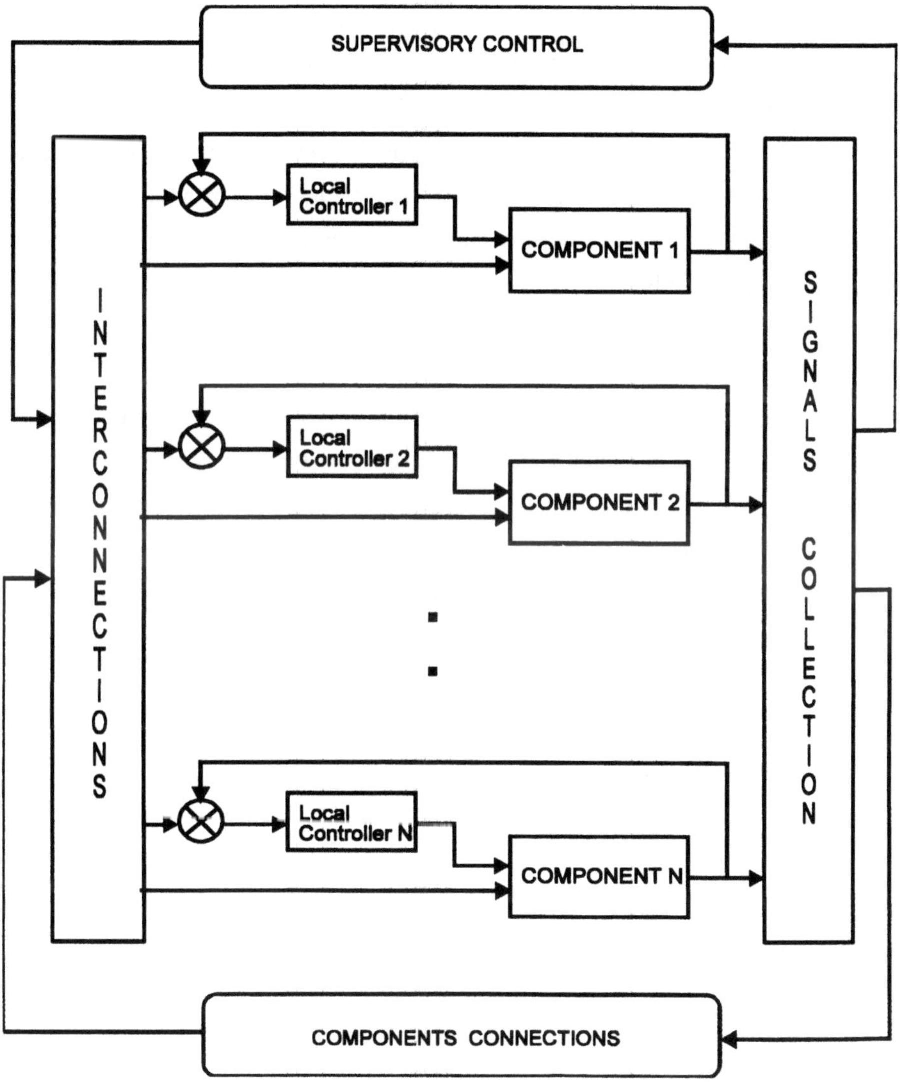

Fig.1.22 A component based block diagram.

The block "Interconnections" on the left hand side of the diagram represents a matrix of 0/1 elements giving all connections between components and allocation of set points coming from the supervisory level to local control loops.

Local controllers are included in the scheme. The set points for local control loops may be constant, if the variable has to be kept on the same level for each operating condition, or may be provided by the supervisor; this applies to the most important variables like load or main steam parameters. Moreover, some adaptive action on local controllers may be performed by the supervisor. It includes tuning controller parameters when the system operating point changes, e.g. the power demand increases, and, consequently, parameters of models used by the supervisor have to be adjusted. The adaptive action is symbolised by a line coming from the Interconnections block to a component controller block.

The block "Signals Collection" is situated on the right hand side of the diagram. This block represents a 0/1 matrix which provides distinction between outputs of the components. Some of the outputs are transmitted to the supervisory level when others constitute interconnections between components. Only the most important variables like electric power, frequency, steam pressure are transmitted to the supervisor.

A similar idea of control block diagrams, although for more simple configuration may be found in [*Hofmann, 1992*].

1.8.5. Conclusion

Two types of block diagrams were presented in this section.

Firstly, the block diagrams representing basic interconnections between different components in CHP and CC schemes were shown in subsections 1.8.1, 1.8.2, 1.8.3. Only simple configurations, including one component of each type were considered. The purpose of these block diagrams was to give an insight into CC/CHP technology, in particular to establish the basic arrangements for the components within the system, i.e. the main energy and mass routes.

Secondly, in subsection 1.8.4 a more general block diagram was presented. The block diagram includes local controllers, interconnections between components and a supervisory action. Moreover, a more complex power plant may be considered as more than one component of any type may appear on the diagram. For example, the plant may consist of two gas turbines, two boilers and one steam turbine, as in [*Nakanomori and Kano, 1992*]. This type of block diagram may be used when designing supervisory control strategies.

1.9 Case Study : Skegton Unit

1.9.1 Introduction

As a case study consider a fictitous CC/CHP configuration called *Skegton Unit* as depicted in Fig. 1.23. The configuration was devised to illustrate the wide variety of control problems which may arise in CC/CHP power plants. The system produces both electricity and steam for external users. The electricity comes from two types of turbines, namely gas turbines and steam turbines, and steam is exported to external users both at high pressure and at low pressure levels. The components of the system include :

- two single shaft gas turbines,
- two boilers - single drum, natural circulation, with an integral deaerator, a superheater, a desuperheater and with an additional firing,
- two condensing, triple stage steam turbines,
- two condensers situated at the outlet of the steam turbines,
- two feedwater systems,
- a gas damper redirecting exhaust gases from two gas turbines,
- steam admission valves, throttle valves and high pressure steam extraction valves for the steam turbines.

The exhaust gases from two gas turbines come to the common gas damper and are redirected to two boilers according to the steam demand. The rest of the exhaust gas is discharged to the atmosphere. Two waste heat recovery boilers with additional firing produce steam, part of which is used by the steam turbines and the rest is exported to external users. Normally, each steam turbine is associated with one boiler, however cross-connections are also possible and may be applied when one of steam turbines is out of order or undergoing maintenance.

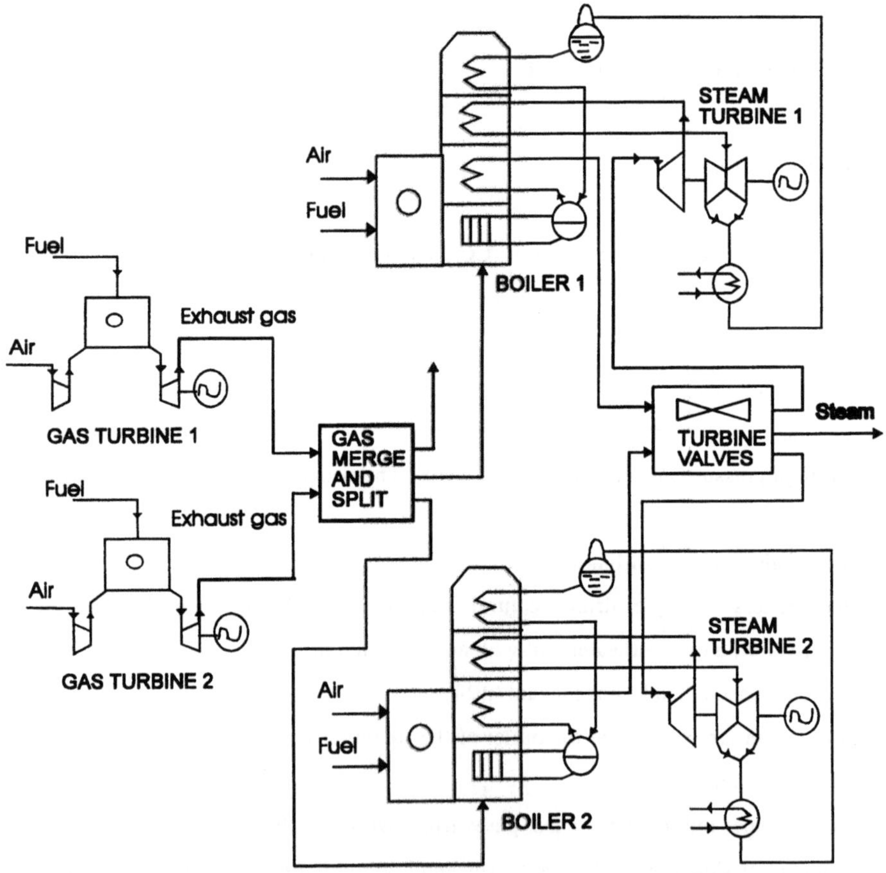

Fig. 1.23 Skegton Unit configuration - schematic diagram.

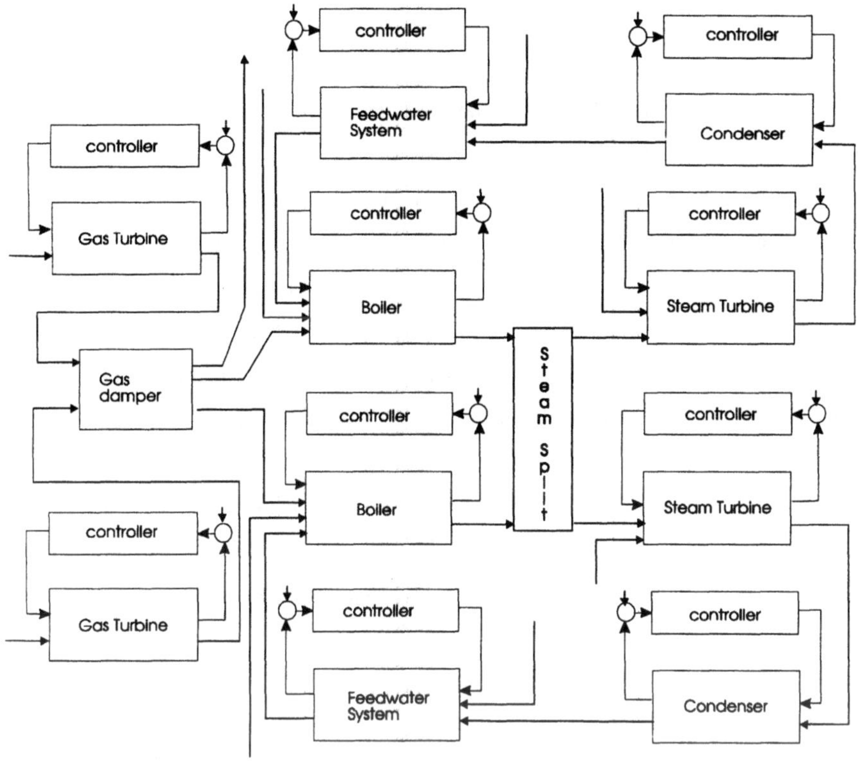

Fig. 1.24 Skegton Unit - generic block diagram.

1.9.2 Block diagrams

The generic block diagram for the system is given in Fig. 1.24. Each major system component is provided with local control loops. Set points for the control loops are supplied by the supervisory control level. The supervisor acts also on the exhaust gas damper and on the steam valves. The purpose of the supervisory actions is to provide an appropriate energy distribution within the system governing the amount of electrical energy produced by the gas and steam turbines and the amount of steam (heat) exported from the system.

1.10 Conclusions

The chapter establishes the main plant components which will be considered in CC/CHP configurations. For each component, a list of possible input and output variables is presented. The list resulted from a literature study on different configurations of low level controllers in power systems. From this general description more specific and simplified models will be constructed. The level of complexity depends on the tasks of supervisory control. Local control loops for power system components are included in the equipment provided by manufacturers of those components. This influences the selection of variables and local controllers for modelling purposes.

The block diagrams presented in section 1.5 give an overall idea of the approach which is proposed. The system is divided into separate blocks, with possible interconnections between blocks and a hierarchical control problem for large scale, nonlinear system with variables constraints is to be examined.

For control purposes three stages of operation have to be distinguished : start-up, shutdown and normal (steady state) operational regime.

Problems, associated with start up and shutdown procedures are especially complicated when the plant structure is complex, including, for example more than one gas turbine associated with one boiler. Presently, those procedures are treated as processes of traversing a series of steady state conditions. Movement from one point to the next is done slowly and carefully as not to involve any transient behaviour. Thus, a better knowledge of dynamic models of the plant together with optimal (minimum time) algorithms could probably increase the speed and efficiency of those operations.

The normal operating regime involves two modes of operation and different control problems associated with each mode.

In the stabilisation mode, when the load demand from the national grid and the heat demand are constant, the main task is to maintain system variables in the presence of disturbances. This regulation is mainly the task of the local control loops. Classical multi-loop designs based on P or PI controllers are standard. However, algorithms based on predictive control methods and self tuning algorithms seem to be suitable for these purposes. For the supervisor, the tasks in steady state conditions involve monitoring safety conditions, collecting the most important measurements and predicting the future behaviour of the system.

When the load associated with the plant changes, the supervisory control level plays crucial role in providing stable and efficient transition to a new steady state conditions. First, new steady state conditions for the whole plant (i.e. mass flows, fuel supply) must be calculated to balance the system and an optimal distribution of load has to be performed. Next, the dynamic problem has to be addressed. The problem is to choose the optimal trajectory for the set points governed by the supervisor leading from actual values to the desired values calculated beforehand as a new optimal steady state conditions. The objective of optimization would be to minimise energy losses and, in the same time, to keep variables within constraints. On the lower level, the problem of optimal trajectories is associated with the task of following a set point trajectory by a local controller.

1.11 References for the Chapter

Ahluvalia K.S. and R. Domenichini ***Dynamic modelling of a combined cycle plant.*** Trans. ASME Journal of Engineering for Gas Turbines and Power, Vol. 112, 164-167, Apr. 1990

Alvarez-Calderon J. ***Operating state-of-the-art computer controlled combined cycle power plant.*** Proc. American Control Conference, 1984

Amendola E., A. Possenti and R. Tarli ***Advanced diagnostic systems for thermoelectric plants operation.*** Proc. IFAC Symposium on Automation and Instrumentation for Power Plants, 1986

Amin M., G.Y. Masada and B.F. Womack ***Application of a multivariable self-tuning controller to a power plant boiler.*** Proc. American Control Conference, 1984

Basu R.N. and L.L. Cogger ***Integrated approach to cogeneration planning, control and management.*** Proc. IFAC Symposium on Automation and Instrumentation for Power Plants, 1986

Boissenin Y. and A. Castanier ***Choosing the right combined cycle power plant.*** Alsthom Gas Turbines Reference Library, 1988

Bolland O. ***A comparative evaluation of advanced combined cycle alternatives.*** Trans. ASME Journal of Engineering for Gas Turbines and Power, vol. 113, 190-197, Apr. 1991

Brown A. (Scottish Power plc.) - ***private communication.*** August 1993

Cegrell T. ***Power System Control Technology.*** Prentice Hall International, 1986

Chiacciarini H.G., A.C. Desages, J.A. Romagnoli and A. Palazoglu ***Variable structure control strategies : application to a MIMO nonlinear steam generating unit.*** Proc. American Control Conference, 1992

Choudhari S.H. ***Start-up and shut-down of large size turbo sets.*** Proc. IFAC Symposium on Automation and Instrumentation for Power Plants, 1986

Cohen H., G.F.C. Rogers and H.I.H. Saravanamuttoo **Gas Turbine Theory (3rd edition).** Longman, 1987

Conradie F.H. and H. Kurten ***Actual dynamic performance of thermal power plants with various designs and control systems.*** Proceedings of the IFAC Symposium on Control of Power Plants and Power Systems, Munich, 1992

Dolbec A.C. ***A high reliability gas turbine control system.*** Proc. American Control Conference, 1984

Domijan Jr. A. and Thompson III D.E. ***Development of a PC-based data acquisition system for performance monitoring on an 81-megawatt steam cycle electric power generating unit.*** IEEE Trans. on Energy Conversion, Vol.6, No.3, 426-431, Sep. 1991

Dwarakanath M.H., B. Dembart, A.M. Erisman, K. Hemmaplardh and J.W. Manke ***A generalized methodology for modelling system components in power system dynamic simulation.*** IEEE Trans. on Power Apparatus and Systems, Vol. PAS-101, No.1, 136-146, Jan.1982

Hofmann A. ***A knowledge-based approach to optimizing and maintaining cogeneration facilities.*** Proceedings of the IFAC Symposium on Control of Power Plants and Power Systems, Munich, 1992

Hogg B.W. and N.M. El-Rabaie ***Generalized predictive control of a steam pressure in a drum boiler.*** IEEE Trans. on Energy Conversion, Vol.5, No.3, 485-492, Sep. 1990

Hogg B.W. and N.M. El-Rabaie ***Multivariable generalized predictive control of a boiler system.*** IEEE Trans. on Energy Conversion, Vol.6, No.2, 282-288, June 1991

Honderd G., P.P.J. van den Bosch, A.C. van der Ham and P. van Dijk ***Stabilization of a 28 MW gas turbine: fault analysis and controller design using specialized CACD tools.*** Proceedings of the IFAC Symposium on Control of Power Plants and Power Systems, Munich, 1992

Hussain A. and H. Seifi ***Dynamic modelling of a single shaft gas turbine.*** Proceedings of the IFAC Symposium on Control of Power Plants and Power Systems, Munich, 1992

IEEE Committee ***Dynamic models for fossil fueled steam units in power system studies.*** IEEE Trans. on Power Systems, Vol.6, No.2, 753-761, May 1991

IEEE Committee ***Dynamic models for steam and hydro turbines in power system studies.*** IEEE Committee Report, IEEE PES Winter Meeting, New York, 1972.

Jericha H. and F. Hoeller ***Combined cycle enhancement.*** Trans. ASME Journal of Engineering for Gas Turbines and Power, Vol. 113, 198-202, Apr. 1991

Kano M. **A combustion air changeover control for heat recovery boilers.** Proceedings of the IFAC Symposium on Control of Power Plants and Power Systems, Munich, 1992

Kapadia M. and C.H. Wolgemuth ***A dynamic model of a condenser in a closed Rankine cycle power plant.*** Proc. American Control Conference, 1984

Keeton J.M. ***Advanced control techniques applied to a fossil power plant.*** Proc. American Control Conference, 1984

Kure-Jensen J. and R. Hanisch ***Integration of steam turbine controls into power plant systems.*** IEEE Trans. on Energy Conversion, Vol.6, No.1, 177-185, March 1991

Mann J. and G.K. Lausterer ***Temperature control using state feedback in a fossil fired power plant.*** Proceedings of the IFAC Symposium on Control of Power Plants and Power Systems, Munich, 1992

Marecki J. ***Combined Heat and Power.*** (in Polish) WNT Warsaw, 1991

Marshall E. ***CHP and deregulation.*** Energy Policy, 73-78, Jan. 1993

de Mello F.P. ***Boiler models for system dynamic performance studies.*** IEEE Trans. on Power Systems, Vol.6, No.1, 66-74, Feb. 1991

Nakanomori K. and M. Kano **On the automatic start-up/shutdown system for a combined cycle with two different gas turbines.** Proceedings of the IFAC Symposium on Control of Power Plants and Power Systems, Munich, 1992

Orchard W.R.H. and P.J. Robinson ***Combined heat and power - the economic advantages of low temperature direct connection systems and their compatibility with existing installations.*** IEE Conference Publication No 192, 1981

Patel L.N. ***Modelling the characteristics of thermal power plant boiler furnaces.*** Proc. IFAC Symposium on Automation and Instrumentation for Power Plants, 1986

Rovnak J.A. and R. Corlis ***Dynamic matrix based control of fossil power plants.*** IEEE Trans. on Energy Conversion, Vol.6, No.2, 320-326, June 1991

Rowen W.I. ***Simplified mathematical representations of heavy-duty gas turbines.*** ASME Journal of Engineering for Power, 865-871, Oct. 1983

Rowen W.I. ***Speedtronic Mark IV control system.*** Alsthom Gas Turbine Reference Library (AGTR 880), 1988

Shields C. ***Current developments in gas turbine combined cycle power plant.*** Publication 453, Institution of Diesel and Gas Turbine Engineers, 1989

Smith L.P., S.S. Godbole, W.C. Howe and J.R. Byrd ***Control system development for atmospheric fluidized-bed combustion.*** Proc. American Control Conference, 1984

Stout T.M. ***An experience with dynamic simulation of an industrial steam system.*** Proc. American Control Conference, 1984

Waddington J. and G.C. Maples ***The control of large coal- and oil- fired generating units.*** Power Engineering Journal, 25-34, Jan.1987

Welfonder E. **Constrained control concepts in power plants and power systems for avoiding emergency conditions.** Proceedings of the IFAC Symposium on Control of Power Plants and Power Systems, Munich, 1992

Wiggin F.T.F. ***Commercial/ operational/ performance effects of adding flexible multi-fuel GT/fired HRB combined cycle to Slough CHP utilities.*** IEE Conference Publication No 192, 1981

Wildi T. ***Electrical Power Technology.*** Wiley, New York, 1985

Yacobucci R.B. ***A control system retrofit for a GE frame 5 turbine/generator unit.*** IEEE Trans. on Energy Conversion, Vol.6, No.2, 225-230, June 1991

Yang P. and B.W. Hogg ***Continuous-time generalized predictive control of a boiler model.*** Proceedings of the IFAC Symposium on Control of Power Plants and Power Systems, Munich, 1992

CHAPTER 2

POWER GENERATION PLANT CONTROL

2.1 Introduction

The recent advances in information technology, increased market competition, the tightening of environmental regulations, the demand for low cost operation and energy efficiency have all influenced the need for new control design philosophies for complex industrial systems. The main features and impact of these changes on the plant-wide control methodologies are summarised below:

- New information technology and computing equipment is becoming progressively faster, more complex and has a shorter life cycle.
- Flexible and distributed plants are increasingly more feasible and demanded in the process industries.
- Total plant optimisation with efficient and reliable unit operation is essential for economic operation.
- The integration of control and instrumentation equipment manufactured by different vendors is a major issue in the control design for complex systems.
- The global coordination of management, operational control and maintenance functions is now an essential part of large scale plant computer control systems.

The design and provision of control systems with the capabilities to integrate a large number of plant functions have been a major concern for control practitioners since the Sixties. The early attempts to produce a working integrated system were concentrated on the application of a central digital computer to plant wide control and the replacement of pneumatic and analogue equipment such as sensors and actuators with their digital or high power electronic counterparts. These new computer-based control systems were known as *Direct Digital Control* (DDC) systems. They were initially developed in the process, petrochemical, power and steel industries [*Williams*, 1964]. The early applications of DDC systems were restricted to small and local plant units. However, this changed in the mid-seventies with the emergence of communication networks which enabled computers to be linked together. This improved communication capability revolutionised the application of DDC systems, leading to the emergence of *Distributed Control Systems* (DCS) and the closely related *Supervisory Control And Data Acquisition* (SCADA) systems. It is perhaps useful to introduce some clarification of this terminology.

Definition 2.1 *Direct Digital Control (DDC)* : A centralised computer system using digital algorithms to replace analogue plant controllers and provide signal processing capabilities.
Remarks :

(i) The term was introduced in the 1960's [*Williams*, 1964].

(ii) The poor integrity of the centralised architecture of DDC's was identified very early in the introduction of this technology

Definition 2.2 *Distributed Control System (DCS)* : An integrated computer network of microprocessor controllers and communications devices used for process control and supervision.
Remarks:

(i) This term arose in the 1970's when computer technology and communications permitted an integrated computer system to be developed.

(ii) The underlying system structure is usually hierarchical and now extends to integrating both distributed control systems with business control systems.

Definition 2.3 *SCADA System* : A supervisory control and data acquisition system using a computer network to control and communicate with, possibly remote, autonomous production systems.

Remarks

(i) The difference between DCS and SCADA is becoming vanishing small as the technological capabilities of DCS and SCADA systems approach each other.

(ii) SCADA is a term generic to certain industrial sectors, for example, the offshore oil and gas industry.

The DCS's were initially used for data collection, limited data processing and sequential control applications. PID controllers were soon incorporated into these real time computer systems. With the new developments in control theory and instrumentation technology, new features were added to such systems. The hierarchical structure of such systems enabled plant supervision and optimisation to be added to direct control. These systems are still under extensive development and form an integral part of any modern manufacturing and processing plant. The potential for advanced control features is slowly being realised and the new methods will exploit the available computing facilities and advanced communication buses. This is leading to a new generation of plant wide control systems which can provide efficient operation, integration, supervision, optimisation, management, maintenance and control. The benefits of such systems include, lower installation and maintenance costs, better system reliability, flexible production configuration and the easier expansion of the plant and the control system.

The review of supervisory control systems in this chapter is mainly technology driven and restricted only to continuous process plants. It is no surprise to find that the innovation in new information technology has long overtaken the theoretical developments in control. While the DCS's are now extensively used in practice, there is no coherent theory for the design and optimisation of these systems. The objective is therefore to firstly review the emerging plant wide control technology and its capabilities. Secondly, industrial examples are used to demonstrate the concepts and the problems within the supervisory control framework. Current control and optimisation theory is then reviewed and some recommendations are established for future possible research directions.

2.2 Some Technological Background

The advances in new computer and information technology has had a major impact on how large scale industrial processes are controlled. The main objective of total plant control is to obtain economic benefits (increases in efficiency, throughput, quality, etc.) by globally integrating the process control functions, the supervisory control modules, information services and management decision making functions. The hardware and software requirements of such a system are determined and constrained by the plant size and geography. Although new plants may be designed to be fully compatible with the new technology, there are still many old plants which require supervisory control design on an individual basis. Most plant wide control systems can be represented by a centralised or a decentralised architecture and these structures are discussed in the following sections.

2.2.1 DCS Architectures

2.2.1.1 Centralised Architecture

The key to this concept is a single computer unit at a location which may be remote from the process under control. This centralised facility executes all the computational and operational tasks required by the process. As well as providing the DDC of the plant, this single computer unit performs all the higher tasks such as real-time data acquisition and processing, archiving, control and monitoring, information analysis and management information. The centralised architecture is shown in Fig. 2.1 [*McKendrick, 1994*]

The early approach to the computer control was through the centralised architecture. However, the complexity and geographical scale of most large scale production processes coupled to the development and falling cost of microprocessor technology with associated advances in communications technology lead to a distributed architecture for supervisory control. Although in any control system, some centralisation is required, this is usually restricted to the top level control functions and tasks. The main disadvantage of full centralisation is lack of flexibility and an inherently low structural reliability. To ensure a minimum downtime, at least two computers should operate in parallel with exactly similar instructions and software. The advantage is that the user should only deal with one type of operating systems and the communication modules are all built in within the computer.

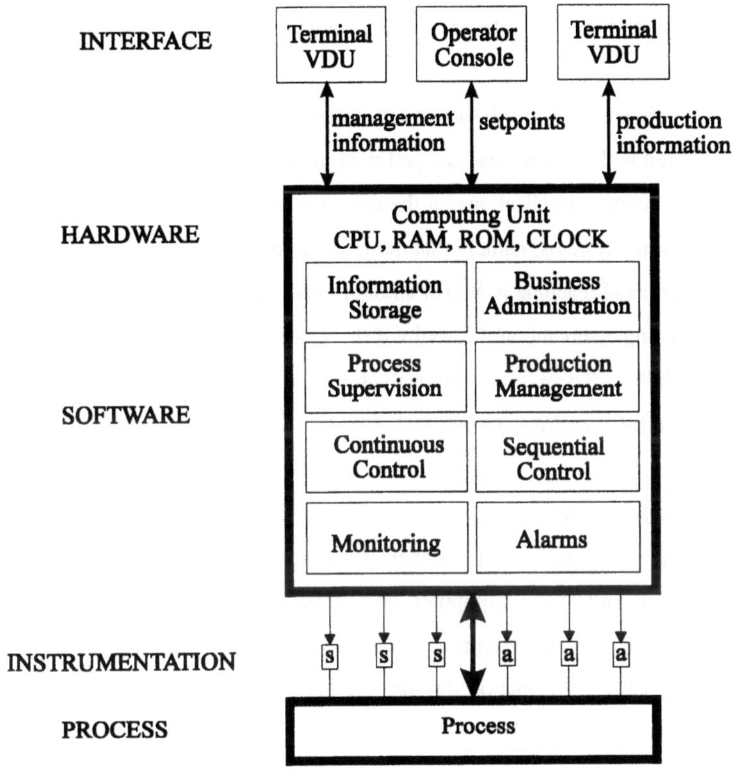

Fig. 2.1 Centralised system architecture for industrial automation

The main components of a centralised system are the interface, the hardware, the software, the control and instrumentation units as shown in Fig. 2.1.

2.2.1.2 The Distributed Architecture

The technological feature of the decentralised or distributed architecture is that the control tasks are spread among a number of computing machines linked by a communications highway. Each control task is a set of control functions. The tasks are usually classified locally and executed on a local computer. The interactions between the tasks are taken into account through higher level modules which initiates, monitors or controls the execution of the lower tasks. Thus, the control tasks are primarily functionally distributed. The physical constraints on the system such as geographical location, safety issues, environmental issues, security, energy utilisation and cost are also determining factors in deciding how the control tasks should be distributed.

To develop an efficient data and information management system, each computing machine should only be fed with the quantity and quality of information needed to executed a designated task. This has certain implications on the design of communication network and the architecture of the control systems, namely that the information flows and links should be established such that each local machine is almost autonomous. This information packaging also introduces a certain implementational flexibility if the global co-ordination is optimised and engineered successfully. There are two distinct forms of distributed architecture: horizontal and hierarchical [*McKendrick, 1994*].

Horizontally Distributed Architecture

In this case the process is partitioned into a number of subprocesses, and each subprocess assigned a local computer unit. Each computer unit is performing some of the tasks associated with the control of the complete process. Global control is achieved by the use of a LAN which transfers essential information through the horizontally distributed computer units as shown in Fig. 2.2.

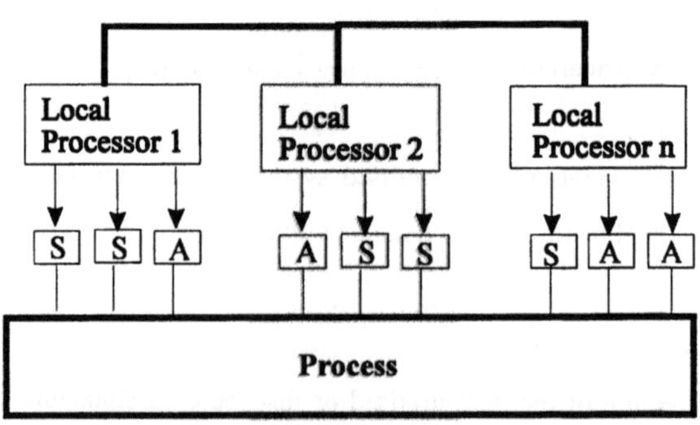

Fig. 2.2 Horizontally distributed system architecture

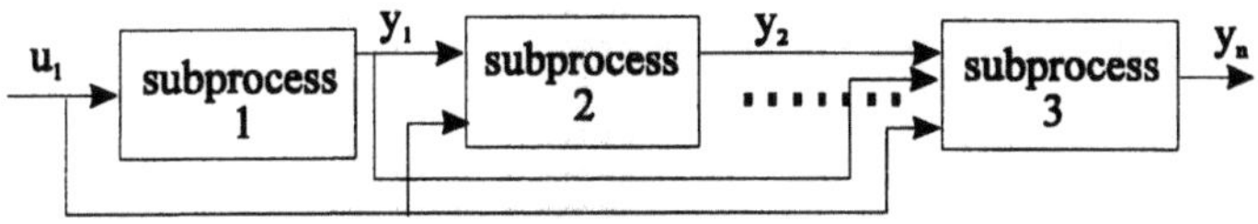

Fig. 2.3 Sequential process plant

This type of distributed control systems is often used for sequential systems where the output of one subprocess is the input to the following subprocess as shown in Fig. 2.3 and where the information is only available at the local subprocess level. This type of plant structure is common in batch processing.

The major drawbacks of a horizontally distributed architecture are:

- It is difficult to globally optimise the overall process due to a lack of information.
- The architecture often duplicates the hardware installed and the software utilised.

Vertical or Hierarchical Distributed Architecture

The vertical or hierarchical distributed architecture is superior to both the centralised and the horizontally distributed architecture. The reason for this is simply that a hierarchical system closely models the dominant characteristics of modern industrial and commercial activities. More precisely:

- Industry is essentially a business and the decision taking mechanisms are essentially hierarchical.
- The quality and quantity of the information needed to take the decisions in controlling all levels of a process are essentially hierarchical.

2.2.2 Distributed Control System Technology

The term *Distributed Control System* is most commonly applied to a plant-wide control system of distributed processors, operational peripherals (VDU's, printers, etc.), minicomputers held together by a communication network operating in real time. A typical structure for a DCS is shown in Fig. 2.4. The basic functions of a DCS can be classified into two groups:

Primary Control Functions: These are related to the direct control of subprocesses at the local level and include feedback control, feedforward control, inferential control, ratio control, cascade control, and so on. The instrumentation used to realise these control functions is called primary instrumentation.

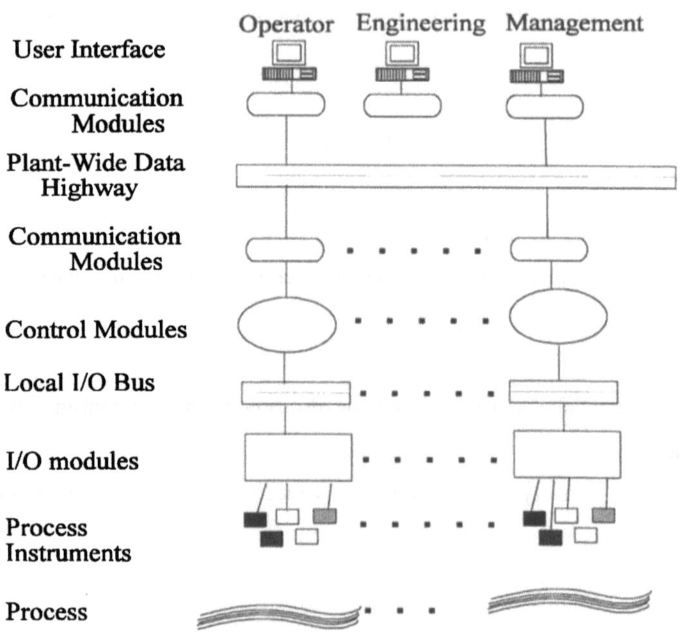

Fig. 2.4 A typical DCS configuration based on functional decomposition

Secondary Control Functions: These are related to the higher levels of the hierarchy and are the supervision, monitoring, management, maintenance and optimisation tasks. Examples are status indicator, alarm, record, optimise, startup, and shutdown routines. The DCS plays a major role in executing the secondary control functions and the associated instrumentation known as secondary instrumentation.

2.2.2.1 Generic Functional Modules

Each module of the architecture shown in Fig. 2.4 is discussed below:

Input/Output Modules provide the main interface between the DCS and the process being controlled. They are used to scan, digitise process instrument input/output signals and provide command inputs to the actuators. They also provide filtering, contact debouncing, low level logic and local alarms. A single input/output board may be connected to a single field device, or a single input/output device to provide loop integrity. A single board may be connected to multiple inputs and multiple outputs.

Typical features of these devices are isolated or non-isolated grounding, a low level of fusing protection, high accuracy and linearity of sampling frequency, protection from electromotive force, an immunity to radio frequency, fail-safe positioning, overload and surge protection, impedance matching, loop feedback sensing, manual override of loop control and a high mean time between failure.

The local *Input/Output Bus* which links the I/O module is a local platform for the communication between the I/O modules and the controller modules. This bus typically operates at a slower speed (0.5 Mbits/sec) than the plant-wide data highway. Communications are provided in various ways, for example polling or scanning of the I/O by the controller modules, serial or parallel communication or both.

Controller modules read, update and calculate new field data. These routines control the subprocesses. The systems PID algorithms are executed by the controller modules. The basic primary and secondary control functions performed by this module can be summarised as: I/O signal characterisation, signal filtering, alarms, control logic, control interlocks, sequencing, trending information, report information and feedback control

Hardware controller modules are microprocessors, minicomputers, PC, PLC, workstation, etc. The key features in selecting a controller hardware modules are memory requirements, idle time, I/O loading, operating system and I/O software capability, the clock cycle and RAM requirement.

Communication modules provide a link between the data highway and the other modules. These are microcomputers or workstations and also known as servers. They basically perform the network management. The features to be considered in their selection are memory requirements, speed and the number of modules to be controlled.

The *Data highway* is a plant wide communication network and different topologies are possible, such as linear, loop or star or a combination of these. The important issues relating to the selection of the data highway are speed of data transmission (1M bits/sec) synchronised versus nonsynchronised communication, token passing versus report by execution, protocol types, data transmission rate and maximum transmission distance.

User interface modules allows the operator, the engineer or manager to interact with the DCS. These devices include display monitors, alphanumeric keyboard, operator keyboard, trackball and mouse, printers and plotters. Recent information technology developments have made operator-interfaces fairly impressive, however, it should be noted that such displays are expensive to produce. Further, the value of the display depends on the accessibility and size of the process database, the integrity of the information, the screen's build or image speed, and the reliability and security of the information.

2.2.2.2 DCS Software

Like any computer system the DCS has to be correctly programmed to perform its tasks in real time. The existing software for DCS's is often nonstandard and proprietary. The basic modules are the operating system, system support software, application software and communication software. These are briefly discussed here:

Operating Systems Software is the executive software for the computer system. Its main distinguishing feature is its real time capability.

System Support Software is used to aid the development of the application programs. These are called system utility programs since they include editors, debuggers, compilers, linkers and so on.

Applications Software is specifically related to the task of configuring a DCS. They usually contain a library of simple operations like reading an input to more complicated modules like a PID or lead/lag algorithm and may extend to complex sequencing, optimisation routines or expert system facilities. These programs are almost exclusively proprietary.

Communication Software is used to establish the communication links between different local computers and the local and highway buses.

It should be noted that the recent trend in the development of DCS systems is toward an open system both in hardware and software. When this happens, it will provide a significant improvement in the flexibility in the DCS system and make the task of designing these systems much easier.

2.2.3 Technological Implications and Potential

2.2.3.1 Economic Benefits

In the 1960's, the pursuit of the economic benefits of global control resulted in the first centralised direct digital control computers and this was reinforced during the 1970's when distributed control systems became widespread. The increasing communication capabilities, data acquisition and processing power of the current and future plant-wide control technology will undoubtedly lead to improved responsiveness in the dynamic scheduling of large complex manufacturing and process plants. The impact of this on the control system will be the closer integration of the control, monitoring and operation of the process units.

This trend can be observed across a wide range of industries but particularly in the large scale industrial processes of the chemical, petroleum, paper and steel sectors. These more recent advances have made supervisory control easier and cheaper to implement and the key benefits are seen as:

Increased Plant Capacity : In terms of the power industry where power generation units are of interest, increased plant capacity is equivalent to higher energy production efficiencies and better availability of the generation units to external loading and demands.

Lower Operating Costs : This has several self explanatory components :

Less fuel feedstock required, improved energy utilisation or production, lower maintenance costs, reduced labour costs, improved plant safety, improved product quality and uniformity, reduced waste, improved process information and management

2.2.3.2 A Classification for Supervisory Control Problems

The complexity and size of large-scale industrial plant, whether it be a thermal power station or a chemical refinery, requires more than loop control or even process unit control. Coordinated total plant-wide control is required for optimised system operation. Supervisory control systems are designed to provide the integration of the industrial subprocesses. There are perhaps two aspects to the design of supervisory control systems:

(i) Responsiveness to external system conditions, for example, reacting to the network loading which impacts the power generation production requirements.

(ii) Responsiveness to internal system conditions, for example optimised system set-up to take account of fault conditions or maintenance schedules. This includes being able to reconfigure the production of power given a particular plant availability scenario.

The supervisory control system design problem often takes the form of on-line optimisation of the primary and secondary control functions. These are classified as:

Scheduling : This is the long term planning problem of operating and directing the plant operation within the constraints of maintenance requirements, faults and machine failures. For example, a balance between power production and loss of plant availability has to be achieved.

Optimal Plant Allocation : This involves the economic balancing of the available throughput against external demand. A typical problem is the selection of unit loading in power generation system from the available units of boilers, gas turbine and steam turbine of different sizes, efficiency and mix of fuel costs. In the short term, optimal allocation has to accommodate unpredictable events like outages, machine failure and similar problems.

Operating Condition Optimisation : This is the problem of establishing the best operating conditions giving maximum efficiency for least economic cost within the primary and secondary system constraints. In combined cycle applications, a more flexible plant availability creates the problem of dynamically transferring plant from one operating condition to another as demanded by the plant allocation schedule.

Unit Optimisation : This is basically achieved by the local loop controller. The feedback control design problem should be solved such that the unit control design specification, formulated within the context of global system requirements, is satisfied. There are numerous techniques in classical and modern control theory to deal with this problem. However, methods dealing with constrained problems are not so common and significant cost benefits will accrue from such techniques.

2.2.3.3 On Designing Supervisory Control Algorithms

An important step in designing a supervisory control systems is the functional specification. This is a description which has been drawn up to answer the following:

- What are the tasks of the supervisor?
- How the tasks are grouped into the primary and secondary control functions?
- How will these tasks be accomplished?
- What resources will be required to implement the supervisor?

Typically the functional specification might consider the following issues:

Regulatory Control Structure : Input/output variables, equations for coupled process variables, control equations, initialisation procedures, fail safe controller strategy, failure mode analysis, operator interfaces and displays.

Monitoring Functions : Condition monitoring, fault monitoring, alarms, interlocks and warning systems.

Performance Calculations : Performance indices, data processing, reporting procedures

Operator Communications : Process data display, operator interface, alarm interface and data representation.

Data base requirements : Data archival, data retrieval, data processing, data management and data security.

Process models and optimisation : Operational objectives, process constraints, process models and optimisation procedures.

Other issues such as implementation, software and hardware selection and economic justification are considered the tasks of project engineers and managers. Traditionally, supervisory control has been considered to be a set point optimisation procedure. Although this is still an important function of supervisory control, it seems prudent to recognise that supervisory systems are now a more substantial technological development component.

A general procedure for the formulation of the optimisation problems occurring in small scale supervisory control has been given by *Seborg et al* (1989) based on earlier work by *Edgar and Himmelblau*, (1988). A more systematic procedure for complex large-scale industrial systems is envisaged. The review which follows is designed to identify the important features of a global design strategy.

2.3 Power Generation Plant Control

Control of power plants is organised in a hierarchy of levels [*Waddington and Maples, 1987*], [*Vahldieck and Krause, 1986*]. The lowest level is that of the actuators. Then the component control level regulates the variables of the given component, like pressures, temperatures, flows, speeds to desired values. Also start-up and shutdown procedures are implemented at that level. The higher levels, which cover the whole power station coordinate the component controls by adjusting desired set points for variables like load and fuel/air ratio. Start-up and shutdown are co-ordinated from these levels. The tasks of these levels include optimization of complete operation by defining desired trajectories and then following them by adjusting set points for lower level. Finally, the managerial level performs long-range optimization, allocates loads to power groups and defines general criteria and constraints (economic dispatch).

2.3.1 Tasks of low level controllers

The low level realises the start-up and shutdown, monitors safety conditions and involves alarm signals with safe shutdown procedures, and performs modulating control on plant variables. The tasks of modulating control [*Peters and deBoer, 1992*], [*Rovnak and Corlis, 1991*], [*Ferrarini and Maffezzoni, 1992*], [*Mann and Lausterer, 1992*] are :

- to stabilise the system variables at desired values,

- to ensure the required dynamic behaviour when changing from one operating point to another.

Some safety demands are incorporated in the modulating control algorithms in the form of variables constraints or additional control loops to stabilise certain parameters, for example :

- the rate of change of boiler wall temperatures when warming up during the start-up procedure must be limited to avoid thermal stresses,
- the rate of change of gas turbine load must be limited to avoid a compressor surge.

Examples of variables which are stabilised because of safety reasons are as follows :

- output steam temperature of the boiler,
- drum water level,
- steam pressure in the boiler (drum steam pressure),
- pressure in the combustion chamber of the boiler (to stabilise the flame),
- inlet gas temperature in the gas turbine,
- amount of polluting substances in exhaust gases from the gas turbine,
- steam pressures inside the steam turbine.

The main control loops for components like steam turbine, boiler, gas turbine, condenser, heat exchanger are described in Chapter 1.

2.3.2 Process variable constraints

Variables constraints may be divided into two groups. The first group are constraints resulting from safety regulations. They are described fully in Chapter 1, section 1.7.3. The most important variables which have to be kept within constraints are listed as :

- turbine exhaust gas temperature,

- turbine vibration,
- lubricant oil temperature,
- lubricant oil pressure,
- generator temperature,
- bearing vibrations,
- valve vibrations,
- steam pressure in the boiler drum,
- superheat or reheat steam temperature,
- drum water level,
- feed pumps and air fans vibrations.

The second group consists of constraints which result from the performance limits or capability of the plant or of the plant components [*Urbancic and Skapin-Rugelj, 1992*]. This group includes :

- electric power produced which is limited by the nominal output power of the installed generators; if the demand is higher than the nominal power and cannot be realised by other stations then some of the users must be cut off,
- heat produced is constrained because of capabilities of heat exchangers - maximum heat flow,
- thermal heat flow in the boiler; the constraint results from combustion parameters like exhaust gas temperature and flow, and also from the size of the boiler,
- input and output steam flows in the steam turbine and other flows in the system are limited by dimensions of pipes,
- heat provided to the heat recovery steam generator in combined cycle results from the power of the gas turbine and the efficiency of the gas cycle,
- parameters of steam or hot water exported outside the plant in Combined Heat and Power are related closely to capabilities of heat exchangers and to properties of piping system used to export heat. However, there are constraints as on the steam or hot water maximum (and minimum) temperature, pressure or flow resulting from agreements with endusers,

- efficiency of energy conversion and efficiency of heat exchange in combined heat and power are limited by the performance of current CC/CHP technology.

2.3.3 Top level control considerations

The functions of the top level (supervisory) control system differ from the low level in their purpose and the complexity of the algorithms used. The supervisory controller performs optimization for the plant trying to minimise global cost and, at the same time, trying to satisfy electricity and heat demand. It involves the following functions :

- energy distribution (e.g. fuel supplied to different components),
- economic optimization - economic load dispatching,
- frequency stabilisation,
- balance of plant equipment usage (i.e. balance of flows),
- tuning for the lower level controllers.

Some considerations concerning the above may be found in [*Basu and Cogger, 1986*], [*Tamura and Hagiwara, 1986*], [*Kure-Jensen and Hanisch, 1991*], [*Waddington and Maples, 1987*]. In [*Moslehi et al., 1991*] a combined heat and power scheme including 19 boilers and 11 steam turbines in four power plants is considered. The model of the system includes the components like steam turbines, boilers, condensers, tanks, deaerators, feedpumps, steam network, electric network. For such a system an optimization in the sense of minimal overall cost is performed by selecting the best dispatch of loads between available components. This is a typical example of a static optimization problem required for the criteria listed above. Certain dynamic functions are also realised by the top level control systems. They include supervising start-up and shutdown operations [*Kano, 1992*] and designing trajectories for load changes in the system. In [*Conradie and Kürten, 1992*] a group of tests is described to check the response capability in the first 15 seconds, illustrating the possible rate of reaction when immediate emergency power is required. The tests are applied to the supervisory control level but the action must be performed at the lower level.

The safety functions realised by top level control system are often connected with outage conditions when, due to big load changes or failure of a part of system, some of electricity consumers must be cut off within a few seconds [*Wildi, 1985*].

2.4 Control for Skegton Unit

2.4.1 Main control problems

The normal operating regime for the case study system involves two modes of operation and different control problems are associated with each mode.

2.4.1.1 Stabilisation mode

The steady state, when achieved, requires the local control loops to stabilise signals in the presence of disturbances. Classical approaches are usually based on P or PI controllers. However, certain simulated or even real applications of modern control techniques to these problems have been reported [*Hogg and El-Rabaie, 1991*], [*Vahldieck and Krause, 1986*]. Algorithms based on long range predictive control methods and self tuning algorithms seem to be suitable for these purposes because they provide the minimisation of output signals variances, can deal with systems with large delays (like boilers) and can accommodate variable constraints.

For the supervisor, the steady state involves monitoring safety conditions, collecting the most important measurements and predicting the future behaviour of the system. Linear and nonlinear estimation techniques may be used for the last objective.

2.4.1.2 Changes of load

When the load associated with the plant changes an extensive reaction from the supervisory level is required. Changes of load may be caused by :

- changes in electric power demand,
- changes in heat demand,
- start-up or shutdown of certain components.

In each case, new steady state conditions for the whole plant (i.e. mass flows, fuel supply) must be calculated to balance the system. Then, an optimal re-distribution of load among available generators (i.e. electrical generators and steam generators) is performed. [*Urbancic and Skapin-Rugelj, 1992*] and [*Moslehi et al., 1991*] report the use of a linear programming method for the optimisation of load distribution.

Next, the dynamic problem may be addressed. The problem is to choose the optimal trajectory for the set points governed by the supervisor leading from the actual values to the desired values calculated beforehand as new optimal steady state conditions. The objective of optimisation would be to minimise energy losses and, at the same time, to keep variables within constraints. For instance, if the load of the gas turbine is to be increased, the increase of the exhaust gas flow must be anticipated. The additional gas flow must not enter the boiler instantaneously as it would cause thermal stresses. Thus, initially, part of the exhaust gas must be directed to bypass the boiler and only gradually the gas flow to the boiler may be increased as the additional firing to the boiler is reduced at the same time. However, the heat from the gas bypassing the boiler cannot be recovered. Consequently, the goal for the supervisor would be to find an optimal trajectory for changing the damper position to minimise losses of energy.

On the lower level, the problem of optimal trajectories is associated with the task of following a set-point trajectory by a local controller. There are several techniques like : model following, pole placement, generalized predictive control which might be used for this purpose.

A task which is shared by both the supervisory level and the low level control is the prediction of future changes in the system variables and using those predictions in control strategies. Because the steam part of the system has very large time constants (~200 seconds) and the sampling time must be rather short (below 1 second) to fulfil safety requirements (e.g. speed drop detection), a very early anticipation of future load changes may improve the quality of control in transient states. Long range predictive controllers might be tested for these tasks.

Separate problems, not mentioned here, are associated with start up and shutdown procedures. Presently, these procedures traverse a series of steady state conditions. Movement from one point to the next is done slowly and carefully as not to involve any transient behaviour [*Terwilliger et al., 1984*]. Thus, a better knowledge of dynamic models of the plant together with optimal (minimum time) algorithms would probably increase the speed and efficiency of these operations.

2.4.2 Local control loops

A more detailed description of local control loops is given in Figs 2.5, 2.6, 2.7.

2.4.2.1 Gas turbine local control loops

In a gas turbine [*Rowen, 1988*], the main control loop adjusts the fuel flow to ensure the correct output power and frequency. In the simplified diagram of Fig. 2.5, the fuel flow is selected as a minimum value of three regulators : power regulator, frequency regulator and exhaust gas temperature regulator. In addition, a lower limit for fuel flow is given to avoid the loss of flame. In normal operating conditions, the PI controller regulating output power is most important. The two remaining controllers are included to avoid exceeding safety constraints for speed and exhaust gas temperature when the load increases. The exhaust gas temperature is controlled by adjusting the input air flow. However, in the Combined Cycle, a proportional controller is used, so that the exhaust gas temperature is kept as high as possible, within permissible limits, to provide maximum energy flow to the boiler. The next loop regulates the NOx level in the exhaust gas by a water spray into the combustion chamber of the turbine. The measurement line delay for chemical analysis is around 15 seconds.

The set points for the output power loop and for the frequency loop must be provided by the supervisor. The set point for the gas temperature loop is constant in normal steady-state conditions, however, it might be required to be adjusted by the supervisor when the load changes.

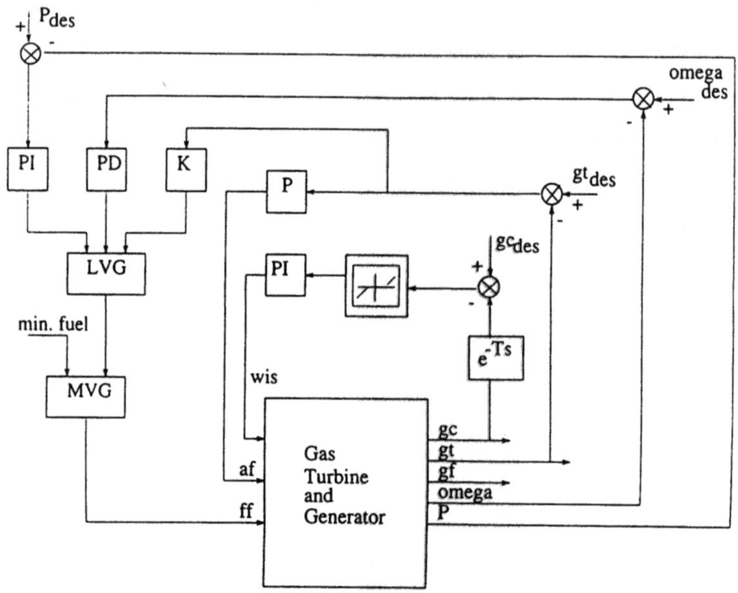

Fig. 2.5 Gas turbine - local control loops.

Main constraints on variables are :

- the output gas temperature, upper and lower constraints depending, on actual fuel flow,
- the frequency and the speed of frequency changes,
- the minimum fuel flow constraint.

The control system illustrated provides safe and sufficiently fast operation using classical P, PI, PD controllers. The strategy is not necessarily the most efficient in the sense of speed and optimal energy usage. The application of more sophisticated controllers might be considered, especially, minimum variance controllers for output gas temperature stabilisation and for maintaining the NO_x level. Long range predictive control may be useful for the main control loop.

2.4.2.2 Boiler local control loops

The boiler local control loops are shown in Fig. 2.6. A simplified boiler model consists of 4 inputs and 4 outputs. The main control loop regulates the fuel flow at the inlet to the boiler. A gas fuel is considered in this example. The fuel flow is regulated by a PI controller using measurements of the output steam pressure. As a set point, the signal related to the desired output power is given. The nonlinear block represents a relation between the power and the desired output steam pressure. In the low load range the boiler is operated at a low pressure level in the constant pressure mode. With increasing load the desired pressure grows in proportion to the desired output power until it achieves a high load constant value [*Welfonder, 1992*]. A parallel feed forward PD control path reduces control effort when the desired power changes. The desired power signal is provided by the supervisor.

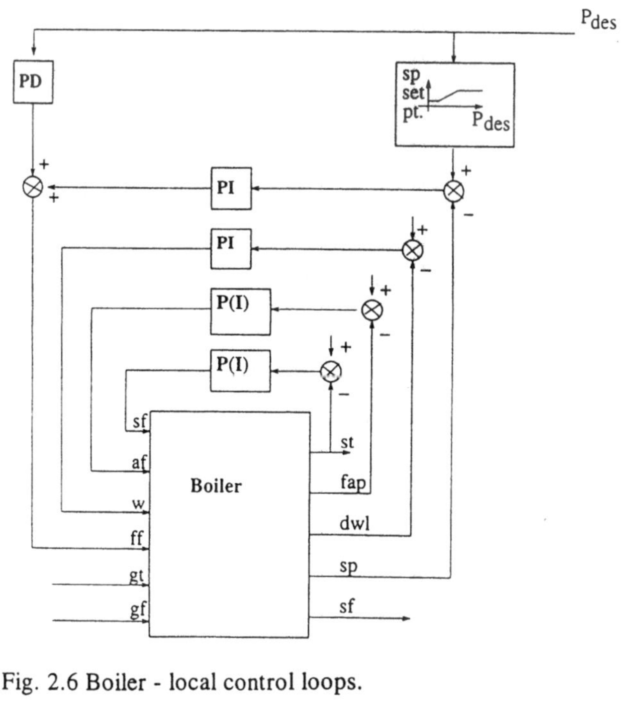

Fig. 2.6 Boiler - local control loops.

The drum water level is a good indicator of the amount of feedwater required by the boiler [*Waddington and Maples, 1987*]. On the diagram, a PI controller uses the difference between the actual and desired drum water levels to regulate the water supply.

The induced air flow to the furnace of the boiler is adjusted to maintain the air pressure in the furnace. A simple P or PI controller may be used.

The outlet steam temperature is stabilised by one or more water sprays to the superheat section of the boiler. In normal operating conditions this loop plays less important role and gives only some fine adjustments [*Welfonder, 1992*]. A simple P or PI controller may be used.

The gas temperature and the gas flow represent interconnection signals from the gas turbine. The gas flow may be adjusted by the supervisor in the gas damper block to provide a proper energy distribution.

The main control problems for the boiler are as follows :

- Keeping all output variables within permissible constraints. This task is realised by a low level stabilisation.
- Providing the rapid change to outlet steam parameters following changes in desired load. The feed forward action on the fuel flow is one solution of the problem. However the speed of operational change has an upper limit caused by possibility of thermal stresses in the boiler or in the steam turbine [*Nakanomori and Kano, 1992*]. Thus, a trajectory following problem arises. The trajectory to be followed is usually a ramp [*Conradie and Kurten, 1992*] and this results from practical experience with the system rather than from simulation and optimisation.
- Maintaining the outlet steam parameters when fuel quality changes. For this purpose, a state controller was proposed in [*Vahldieck and Krause, 1986*] as being more effective than the classical PI/PD cascade.

Additional problems arise when the load of the gas turbine changes causing rapid changes in the gas flow. Immediate action must be taken to avoid exceeding temperature and pressure constraints for the outlet steam from the boiler and to prevent thermal stresses. The closed loop control on the fuel flow to the boiler is not satisfactorily in that case because the boiler response is slow. The solution is to control a damper position in the gas damper. This must be done by the supervisor via static optimisation of energy distribution and dynamic optimisation i.e. choosing the best trajectory for the transient period.

2.4.2.3 Steam turbine local control loops

The steam turbine local control loops are presented in Fig. 2.7. Both steam turbines work in the boiler following mode, so that the load and frequency signals adjust admission valves to the steam turbine. Frequency stabilisation via a proportional controller forms a faster, inner loop. Parallel to it, is a power loop using integral control. In addition, constraints on input steam pressure are taken into account. They are represented by a nonlinear relation between desired power and desired steam pressure as described in 2.4.2.2. Thus, if the steam pressure approaches a permissible limit, an additional action is taken to move the steam pressure back to normal conditions by opening or closing the bypass valves.

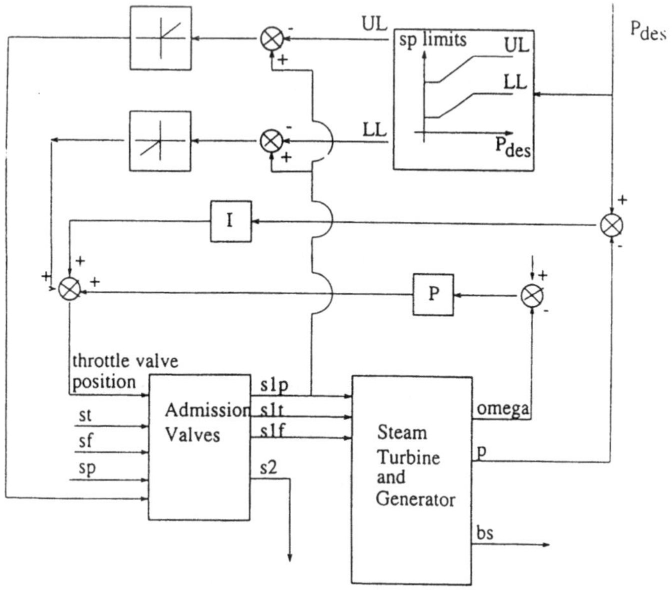

Fig. 2.7 Steam turbine - local control loops.

The main control problems for the steam turbine lie in the need for fast response on load changes without causing thermal stresses, overspeeding or large speed drops. At the same time a proper adjustment of admission valves and steam extraction valves (redirecting the steam outside the system) plays crucial role in energy distribution, especially when the turbine works in a CHP configuration. This must be done by the supervisor using simulation experience and a static optimisation approach.

2.5 Supervisory Control Problems Summarised

Most of the current textbooks on power generation systems provide only a very simple model of the turbine-generator component and concentrate on the network problems, that is control of frequency, and load. Very few venture downstream beyond the steam valve to consider the full complexity of the power-station process control problem. For the case of combined cycle or co-generation plant this approach is inappropriate. The long term planning for co-generation plant is a complicated issue which attempts to resolve the following:

- (i) The anticipated future load demands, and frequency stabilisation.
- (ii) The size and type of plant mix to deliver these loads.
- (iii) The minimum capital and operational costs over the plant lifetime.
- (iv) The required security and reliability of supply.
- (v) The degree of integration of the co-generation plant with enduser heat and power plant and processes.

Solutions of this problem have been proposed using linear programming methods, [*Norman and Rubbers*, 1980] for example. [*Basu and Cogger*, 1986] have strongly argued that simultaneously with solving the problem of optimum plant mix, it is necessary to develop the load management strategy and verify the strategy using design simulation tools. Part of their argument is based on the premise that cost benefits will accrue from good management of all the resources of the installation. But another part of their argument depends on the optimal and integrated operation of co-generation plant and the potential for advanced control technology to achieve this. For example, [*Basu and Cogger,* 1986] cite boiler control optimization in response to external electric and heat load demands as important for cost savings in co-generation plant.

In this section, several technical areas are identified since together they appear to form a framework for control studies with the supervisory problems of combined heat and power for co-generation plant. Each technical area is motivated by one or more published papers and a conclusion is drawn from the study.

2.5.1 Load Management Strategies

Since load management is primarily about reducing costs (and thereby increasing profits) there is a very extensive literature available. The most obvious feature of this literature is the complexity and the particularity of load management problems, coupled with the widespread use of linear programming techniques to solve the resulting optimization problems [*Stocks et al, 1985 ; Moslehi et al, 1990 ; Urbancic and Skapin-Rugelj, 1992 ; Hofman, 1992*].

Moslehi et al examines the problem of optimising the operation of a network of combined heat and power plants. The example pursued comprised four CHP plant totalling 11 turbine-generators and 19 boilers integrated with a large-scale petro-chemical plant, and a neighbouring electrical utility grid. The optimised scheduling showed overall monetary savings of 1% of the total energy budget. The approach of this particular paper looks as though it has the potential to describe and optimise the scheduling of different plant types within one station.

The paper by Urbancic and Skapin-Rugelj (1992) tackles a smaller scale combined heat and power problem which has some special features. The configuration was a single CHP unit which supplies a space heating scheme. The problem was to find the optimal daily operation of the CHP plant and the optimised heat used by the space heating network. Two types of thermal heat stores were assumed and a linear programming problem was solved. The outcome was a heat/electric power generation profile for a 24 hour period. Optimization of the switching strategy for energy production was also deemed feasible.

Hofman (1992) reported similar work but concentrated on an Expert System as the framework for the regulatory and scheduling problems. Although the G2 shell was used, the central calculation again involved a Linear Programming problem.

Conclusion 2.1 : *The economical optimised operation of combined heat and power plant requires facilities to implement dynamic load management strategies*

2.5.2 Performance Optimization

Even within a dynamic load management scheme there will be many period of quasi-steady state operation. It is highly desirable that during these periods units operate at optimised setpoint conditions. Bhandari et al (1986) have described the value of determining indices for monitoring the control performance of individual process units. The potential of the information technology in new DCS's could be used for this purpose. Non-standard performance indices were devised for distillation by Bhandari et al and online computation was shown to be invaluable for the following purposes :

(i) *Process monitoring and optimization :* By tracking the performance index over a period of time, operators can nudge the process to optimal performance using online adjustments.

(ii) *Trouble shooting :* A drifting value of performance index can be giving an early warning of process faults, for example, leaks, process fouling or other fault conditions.

(iii) *Learning about process behaviour* : A meaningful performance index could be used to validate steady-state or dynamic process models. *Bhandari et al* claim they might also be used to establish cause and effect relationships important in understanding processes.

(iv) *Management tools* : The methods of quality control are about continual vigilance as much as they are about individual measurements. Nevertheless, any measure like a performance index which concentrates attention on quality can make a valuable contribution to management practice.

(v) *Further technological development* : As a consequence of increased knowledge about process operation obtained via a performance index, there are often indications of how to use new controllers or other technological innovations to improve performance.

The construction of economically and/or technically valid performance indices allows the identification of optimised process unit setpoints for minimal cost performance. *Dieck-Assad et al* (1987) provide one demonstration of how a physically realistic performance index is used to determine optimal setpoint conditions for given external loadings. These optimal conditions are obtained for the setpoint generators functions used by the electronic controllers. In effect a constrained finite dimensional nonlinear optimization problem was solved to obtain the optimal setpoints. This particular study was supported by field trial tests and the financial cost savings (.28 cents/KWh) estimated for each load level optimised.

Conclusion 2.2 : *For given external loading profiles, different load allocations will occur within the power station. For economic and energy efficient operation it is necessary to determine the optimised process unit setpoints.*

2.5.3 Dynamic Set-Point Manoeuvres

In conventional power plant engineering, external load changes were often absorbed by using designated peak shaving facilities. For example, the Japanese solution was to use nuclear plant for base-load, thermal plant for the large slow load changes and hydro-electric plant was used to cope with rapid load changes. As the Japanese requirement for power grew, the hydro-electric capacity became insufficient for peak-shaving and thermal plant had to be used. This meant that the thermal plant operator had to develop control schemes for managing the more rapidly changing load requirements. It was this situation which lead

Uchida and his colleagues (1981, 1983) to investigate novel and advanced control strategies to bring increased operational flexibility to the speed of response of thermal plant.

The situation described by *Uchida et al* (1981) for boiler-turbine control requires a load demand change to be met by adjusting the steam flow rates as quickly as possible. However to maintain plant efficiency and prolong active plant life the steam temperature should be kept to within ± 1% of nominal (583°C). The plant configuration is that shown in Fig. 2.8.

Note how the system is treated with its proprietary PID controllers in place. Consequently, the reference settings for fuel flow, F_F^D for spray flowrate, F_{SP}^D and for gas damper opening, G_D^D have to be manipulated to maintain the changed superheater, and reheat steam output flowrates without violating the steam temperature constraints:

$$\left|T_{SH}^e(t)\right| = \left|T_{SH}^R - T_{SH}(t)\right| \le \delta_1$$

and

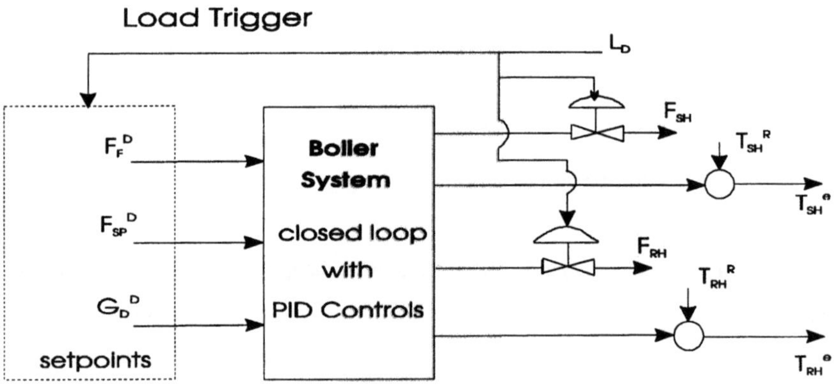

Fig. 2.8 : Closed Loop Boiler System. Key : *Controls (Reference Values) : Fuel Flow,* F_F^D, *Spray Flow,* F_{SP}^D , *Gas Damper,* G_D^D ; *Outputs : Steam Flows,* F_{SH}, F_{RH} , *Superheater Temperature,* T_{SH} , *Reheater Temp,* T_{RH} ; *External Input : Load demand* L_D.

$$\left|T_{RH}^{e}(t)\right| = \left|T_{RH}^{R} - T_{RH}(t)\right| \le \delta_2$$

where δ_1 and δ_2 are 1% of the superheater and reheater steam reference temperature respectively.

If the system equation is written for the boiler of Fig. 2.8, then the following obtains:

$$\begin{bmatrix} T_{SH}^{e}(s) \\ T_{RH}^{e}(s) \end{bmatrix} = \begin{bmatrix} G_{11}(s) & G_{12}(s) & G_{13}(s) \\ G_{21}(s) & G_{22}(s) & G_{23}(s) \\ G_{31}(s) & G_{32}(s) & G_{33}(s) \end{bmatrix} \begin{bmatrix} F_{F}^{D}(s) \\ F_{SP}^{D}(s) \\ G_{D}^{D}(s) \end{bmatrix} + \begin{bmatrix} C_1(s) \\ C_2(s) \end{bmatrix} D_D(s)$$

The solution proposed by *Uchida et al* (1981) has two components (i) feedforward action to eliminate the effect of the external load change and (ii) feedback using the error signals $T_{SH}^{e}(t)$ and $T_{RH}^{e}(t)$ to improve performance.

The interesting feature of the feedforward problem is that the new reference levels for F_F^D D_{SP}^E and G_D^D are sought to maintain the demanded steam flow to the turbine without appreciably disturbing the output steam temperatures, $T_{SH}(t)$ and $T_{RH}(t)$ from their respective setpoints. This is a constrained control problem and a novel solution is proposed and implemented by *Uchida et al.*, (*1981*). This involved a Linear Programming problem to determine a sequence of feasible future setpoints. Using this future setpoint trajectory the boiler system was moved from one operating point to another. The procedure was extended to incorporate two other features : (i) ramp-like load demand changes and (ii) load dependent boiler dynamics (that is, the procedure was made adaptive).

Other authors have also dealt with the need to incorporate system constraints into operational manoeuvres. *Nakanomori and Kano* (1992) report the development of procedures for start-up and shut-down and total load control for a combined cycle plant. Although little analysis is presented, the paper is notable for the hardware and software hierarchies described. The authors highlight the important issues during the manoeuvres namely the operational constraints arising from thermal stress and the need to optimise the rate at which the load demand can be met. Another key point to note from this paper is the way hierarchical concepts permeate the control system and sequencing structure.

Conclusion 2.3 : *In common with many process control applications it is not possible to move from one power plant operating condition to another instantaneously. Previous ad-hoc methods to automate the transition usually involves ramping the setpoint changes to be made over some pre-determined time horizon [Tsai et al., 1986]. Clearly two conclusions arise:*

(i) *Optimal methods for the selection of future setpoint trajectories would be a fruitful research issue.*

(ii) *Once the future set-point trajectory is available, control methods which use this information should be selected for the control of the process.*

2.5.4 Low Level Control Structures

In 1982, *Bransby* reported on the state of the art for CEGB power station control. Direct digital control was the current hardware methodology and *CUTLASS* software was used for implementation. At that time the basic multivariable nature of the boiler-turbine had been recognised. The prevailing control schemes were generally boiler follows turbine and turbine follows boiler modes. However, more complicated multivariable schemes were beginning to emerge, for example, a high security control scheme (*Bransby*, 1992).

It is interesting to compare the 1982 status of low level control with that prevailing a decade later. *Welfonder* (1992) shows a recent steam power plant control system which indicates that adaptation has been engineered by the use of nonlinear gain scheduling characteristics, see Fig. 2.9. The same figure shows the intensive use of multiloop feedback, cascade control loops and feedforward loops. The control structure is strongly motivated by a physical understanding of the system. Controllers are essentially classical, P, PD, PI with the appropriate jacketting software for anti-integral windup, derivative kick protection and bumpless transfer. A key consideration of the *Welfonder* paper is constrained control to avoid emergency conditions.

Conclusion 3.4 : ***At the low level control layer in the hierarchy, the control structure has the following features:***

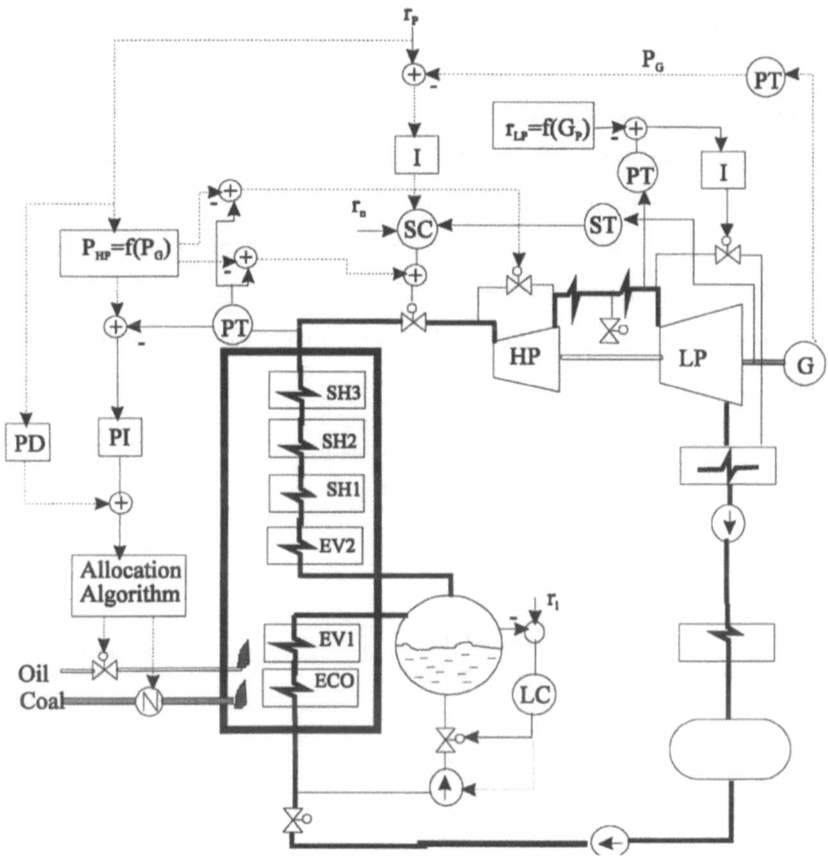

Fig. 2.9 A power plant control configuration [*Welfonder, 1992*]

(i) *it is multiloop and with classical controllers*

(ii) *it is gain scheduled and has feedforward to assist with the disturbance rejection*

(iii) *much of the complexity is designed to achieve inherent constrained operation*

2.5.5 Conclusions

Significant research activities exist in the planning, optimization and supervisory aspects of combined heat and power or co-generation systems. The concept of a hierarchy is pervasive in the hardware, software and the control for these systems. The value of a dynamic simulation to clarify feasible load management strategies, dynamic setpoint manoeuvres and low level control effectiveness seems self-evident from the literature surveyed. The hierarchical diagram of Fig. 2.10 shows how the various supervisory control concepts are related.

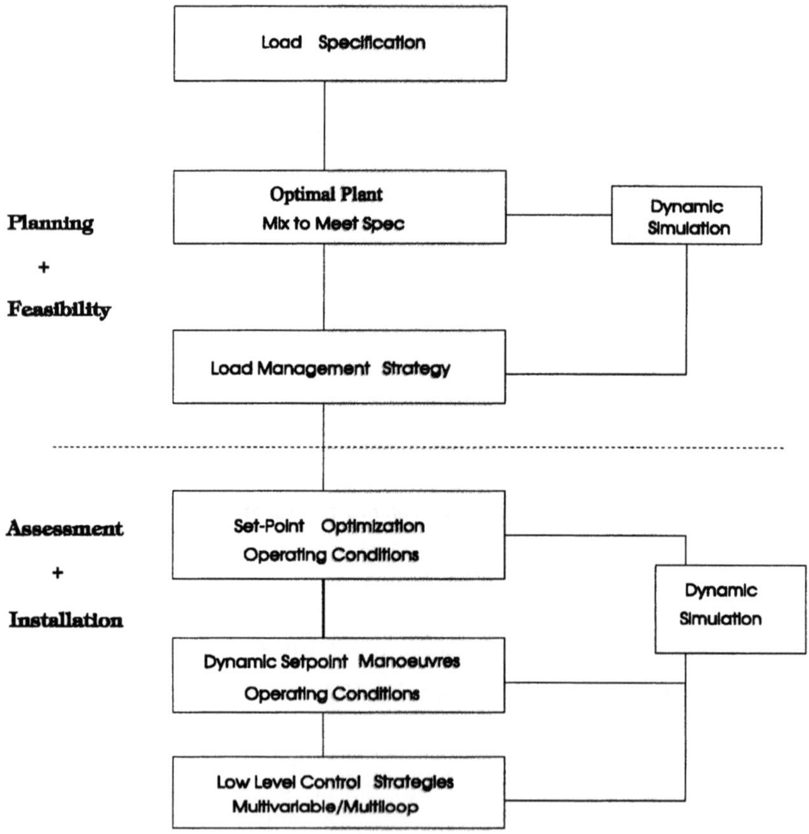

Fig. 2.10 Power plant supervisory control concepts.

2.6 Chapter Conclusions

A systematic review of some areas important to the future work programme and research for power generation control systems for combined-cycle and combined heat and power installations was presented in this chapter. The remit of the review was to identify valuable generic research problems and solution directions.

The chapter progressed by looking at current distributed control system technology, and pursued a detailed structured review of power generation control problems. The main observations are :

DCS Technology :

Computing technology's relentless progress and increasing computational speed, communications sophistication and information presentation complexity provides real opportunities for new supervisory and feedback controls strategies. However, methods which provide a global solution to plant-wide control which have high integrity, robustness, simplicity and can demonstrate economic effectiveness are sought.

Power Generation Control Problems:

In combined-cycle and combined heat and power installations reliable flexible operation providing overall economic benefits in minimising daily costs and maximising plant life-times is sought. A hierarchical sequence of control problems was discussed.

Control Research :

Control theory continues to develop although the preoccupations of the theorists do not always correspond to pressing industrial control needs and objectives. By examining the structure of power generation control problems some useful generic research avenues have been identified.

2.6.1 Recommendations

The recommendations are derived from the practical considerations of this chapter and the classes of problems identified are :

1. *Load Management Strategy*

   This area provides the economic and operational framework for supervisory control design.

2. *Optimised Operation*

   A steady-state and dynamic set-point determination algorithm using various types of performance indices is necessary to implement the load management strategy.

3. *Dynamic set-point manoeuvres*

   A real-time control algorithms for altering the operational points of process units whilst keeping within operational constraints should be sought.

4. *Multiloop control design*

   The practical emphasis on multiloop, classical PID controllers should be met by a reliable and robust control design technique based on such controllers.

2.7 References for the Chapter

Basu R.N. and L.L. Cogger ***Integrated approach to cogeneration planning, control and management.*** Proc. IFAC Symposium on Automation and Instrumentation for Power Plants, 1986

Bhandari, V.A., R. Paradis and A.C. Saxena ***Using performance indices for better control***, Source unknown, ca 1986.

Bransby M.L ***Direct digital control in CEGB Power Stations***, Chapter 9, Eds. Bennet S., and D.A. Linkens Computer Control of Industrial Processes, Peter Peregrinus Ltd. Stevenage, UK, ISBN 0-90604880X, pp. 155-169., 1982.

Conradie F.H. and H. Kurten ***Actual dynamic performance of thermal power plants with various designs and control systems.*** Proceedings of the IFAC Symposium on Control of Power Plants and Power Systems, Munich, 1992

Dieck-Assad, G., G.Y. Masada and R.H. Flake ***Optimal set-point scheduling in a boiler-turbine systems,*** IEEE Trans. Energy Conversion, Vol. 2, No. 3, pp 388-395, 1987.

Edgar, T.F., and D.M. Himmelblau ***Optimization of chemical processes***, McGraw Hill, New York, 1988.

Ferrarini L. and C. Maffezzoni ***A CAD environment for the design of power plant control systems.*** Proceedings of the IFAC Symposium on Control of Power Plants and Power Systems, Munich, 1992

Hofman, A ***A knowledge-based approach to optimizing and maintaining co-generation facilities,*** Proc. IFAC. Symp. Control of Power Plants and Power Systems, Munich, Germany, pp. 211-216., 1992.

Hogg B.W. and N.M. El-Rabaie ***Multivariable generalized predictive control of a boiler system.*** IEEE Trans. on Energy Conversion, Vol.6, No.2, 282-288, June 1991

Kano M. ***A combustion air changeover control for heat recovery boilers.*** Proceedings of the IFAC Symposium on Control of Power Plants and Power Systems, Munich, 1992

Kure-Jensen J. and R. Hanisch ***Integration of steam turbine controls into power plant systems.*** IEEE Trans. on Energy Conversion, Vol.6, No.1, 177-185, March 1991

Mann J. and G.K. Lausterer ***Temperature control using state feedback in a fossil fired power plant.*** Proceedings of the IFAC Symposium on Control of Power Plants and Power Systems, Munich, 1992

McKendrick J.A., M.Phil Thesis, Industrial Control Centre, University of Strathclyde, Glasgow, 1994 (in print).

Moslehi K., M. Khadem, R. Bernal and G. Hernandez ***Optimization of multiplant cogeneration system operation including electric and steam networks.*** IEEE Trans. on Power Systems, Vol.6, No.2, 484-490, May 1991

Nakanomori K. and M. Kano ***On the automatic start-up/shutdown system for a combined cycle with two different gas turbines.*** Proceedings of the IFAC Symposium on Control of Power Plants and Power Systems, Munich, 1992

Norman, H.B. and P.J.E.Rubbers ***A generation expansion program to achieve optimum plant mix and minimum cost of a fixed reliability of supply***, Proc. IFAC Symp. Autom. Control in Power Generation, Distribution and Protection, pp 553-558, 1980.

Peters S.A.W.M. and W.W. de Boer ***Optimal output feedback for the optimization of a conventional steam temperature controller.*** Proceedings of the IFAC Symposium on Control of Power Plants and Power Systems, Munich, 1992

Rovnak J.A. and R. Corlis ***Dynamic matrix based control of fossil power plants.*** IEEE Trans. on Energy Conversion, Vol.6, No.2, 320-326, June 1991

Rowen W.I. ***Speedtronic Mark IV control system.*** Alsthom Gas Turbine Reference Library (AGTR 880), 1988

Seborg, D.E., T.F. Edgar and D.A. Mellichamp ***Process dynamics and control***, John Wiley, New York., 1989.

Stocks, K.J., K.J. Maher, D. Le, and C.H. Bannister ***A linear programming model for assessing co-generation strategies OMEGA***, Int.. J. Mgmt. Sci., Vol. 13, pp 541-554, 1985.

Tamura Y. and S. Hagiwara ***A new algorithm of on-line ELD for thermal power plants.*** Proc. IFAC Symposium on Automation and Instrumentation for Power Plants, 1986

Terwilliger G.E., A.S. Brower, R.S. Baheti, R.E. Smith and G.H. Quentin ***Advanced control aspects of the IGCC concept.*** Proc. American Control Conference, 1984

Tsai, T.H., J.W. Lane and C.S. Lin ***Modern control techniques for the process industries*** Marcel Dekker Inc., New York ISBN 0-8247-7549-X, 1986.

Uchida M., N. Nakamura and K.Kawai ***Application of linear programming to thermal power plant control,*** IFAC 8th Triennial World Congress, Vol. XX, Paper 97.12, Kyoto, Japan, 1981.

Uchida M. and H. Nakamura ***Optimal control for thermal power plant in Kyushu Electric Power Company***, IFAC Workshop on Modeling and Control of Electric Power Plants, 1983.

Urbancic A. and M. Skapin-Rugelj ***Optimization of daily combined heat and power production.*** Proceedings of the IFAC Symposium on Control of Power Plants and Power Systems, Munich, 1992

Vahldieck R. and H. Krause ***Practical experience with progressive automation concepts in power plants.*** Proc. IFAC Symposium on Automation and Instrumentation for Power Plants, 1986

Waddington J. and G.C. Maples ***The control of large coal- and oil- fired generating units.*** Power Engineering Journal, 25-34, Jan.1987

Welfonder E. ***Constrained control concepts in power plants and power systems for avoiding emergency conditions.*** Proceedings of the IFAC Symposium on Control of Power Plants and Power Systems, Munich, 1992

Wildi T. ***Electrical Power Technology.*** Wiley, New York, 1985

Williams. T.J. ***Direct digital control computers - A coming revolution in process control***, Proc. Texas A.and M. 19th Annual Symposium Instrumentation for the Process Industries, (70-81), College Station, Texas, USA., 1964.

CHAPTER 3

SIMULATION TOOLS

3.1 Introduction

System simulation really has changed! In the past the most difficult part of any control design or system feasibility study was developing a system simulation. Ten or twelve years ago, the system engineer was faced with directly using a high level language like FORTRAN, supported by libraries of numerical algorithms. Today, Windows, thc mouse and advanced computer graphics reign supreme. The second surprising feature of todays simulation software is the range of system types which may be easily simulated. It is perhaps useful to start from this point and investgate the types of systems which can be simulated.

3.2 Engineering System Types

Industrial systems of interest to the engineer may be divided into three categories:

- Discrete systems : These are sequential processes of individual operations as might be found in a factory. A typical example is a packaging line comprising a series of conveyor belt linked stations. Their sequential mode of operation produces discrete event lists like: *position box, seal lid, apply label, polythene wrap, stack in fours.*

- Process plant : These are large scale industrial processes operating in continuous or batch mode. The material and energy flows are essentially continuous-time, for example, powders, liquids or gases.
- General dynamical systems : This is the very large class of technological systems exemplified by a jet aircraft, an automobile engine, a CD player, a ship propulsion system or a process unit. Such general engineering systems range from the very small (an electronic circuit) to the very large (a power station).

This might seem an unusual division of system types involved in industrial control and control engineering, however, the thirteen or more simulation tools surveyed for this chapter are readily discussed using these headings. One finding emerging from the survey was that some simulation tools have a surprising versatility and potential which is only just being realised. Indeed several small, and some not so small, companies have managed to carve out a market based on teaching customers how to use these advanced system simulation tools in novel ways on some very diverse projects. In other industrial sectors, it was found that some large companies have standardised on one advanced system simulation tool, and other companies in the sector were following this market leader. One implication of this practice was that subcontractors have to produce system simulations using the favoured simulation software. To survey the wide availability of new simulation software, the three system categories above proved to be very useful.

3.2.1 Discrete System Simulation

Discrete systems originate in factory, production and manufacturing processes. These systems comprise many interlocking flows and networks of manufactured components which have to be assembled, packed and stored. For competitive efficiency, production lines have to be optimised and bottle-necks eliminated. These are the processes of the Computer Integrated Manufacturing (CIM) technology area. However, it is not a good idea to be too restrictive about industrial sectors since one vendor applied discrete systems simulation software to the real-time optimization of hospital throughput with considerable success!

Typical discrete system software tools are SIMAN/CINEMA, SIMPLE++, SIMFACTORY/SIMPROCESS, and WITNESS. They provide animated graphics to demonstrate material flows, process scheduling, product assembly strategies, factory logistics, warehousing requirements, automated guidance vehicle utilization and so on. Apart from the value of the visualizing factory/production line operation, quantitative results are obtained from the simulation runs in tabular, or graphical form, for example pie charts or histograms or through statistical analysis.

What is particularly impressive about these particular similation tools is the way very complex system simulations can be constructed using hierarchical system build utilities. And further, the fact that the system constructed can be of a very general nature so that questions like staff allocation in factory operations can be investigated. It might be difficult to formalise and solve these problems by any other method than a simulation. Consequently, the emphasis is on optimization and understanding the system by direct simulation. This means formulating ideas about the system and running the simulation to see if they are successful since these packages seem to have few formal or theoretical tools available. The main reason for this is that theory of discrete processes is still in its infancy and only now are researchers like Ho, Wonham and Ramadge producing the first formal results in this area [*Ho, 1989; Ramadge and Wonham, 1987*].

3.2.2 Process Control Systems

The category of process plant simulations recognises that some specialist tools have been devised for the process engineering industry. Primarily the industrial sector is that of petrochemicals, pharmaceuticals, biotechnology, mineral processing, gas processing and so on. Process plant design is generally via the chemical and process engineers flowsheet. Typical features include steady-state design using material and process heat balances and interfaces to physical property databases (sometimes the PPD's are of a proprietary nature). What make such tools distinctive is a chemical engineering library containing modules like heat exchangers, distillation columns, reboilers and chemical reactors.

Icon based process control tools are state of the art for this sector. For example, there is the Norwegian product CADAS (Computer Aided Design Analyse and Synthesis of industrial processes). The CADAS simulation is configured through a graphical user interface with unit process models and control/logic function blocks being selected from a library and inserted into a flowsheet. Signal and process lines are then connected at will. Another comprehensive group of packages for the process industry is ASPENPLUS, MAX and SPEEDUP. Again the emphasis of these packages is for chemical and process engineering problems. Features of ASPENPLUS include an extensive library of unit operations with tailored numerical and optimization procedures. This package also extends to the economic modelling of processes. The MAX package is personal computer process simulation package which is graphically orientated using the mouse and a unit library. From the same stable is SPEEDUP which a dynamical process simulation tool. The dynamical dimension to process simulation is essential to resolve on-line operational problems. Features like an extensive library, user specified model facilities, and a specialist mathematical library make SPEEDUP a well-known example of a dynamic process plant software in this particular system area.

3.2.3 General Dynamical Systems

The detailed discussion in this chapter concentrates on software for the category of general dynamical systems. The reasons for this are two-fold. Firstly, the tools discussed are general purpose, and have the potential to simulate a wide range technological processes. For example, known applications range from the dynamics of a cold rolling mill, to the operational procedure and dynamics for a gas turbine, down to simple nonlinear system simulations like a demonstration of the Astrom and Hagglund relay concept which is at the heart of the autotune PID method. Libraries of unit and component models for specific industrial sectors are now being developed for many of these tools.

The second, and quite influential reason, is that these tools are becoming commonplace on many engineering courses world-wide. The ease of clicking together several process units or engineering components and immediately producing a system response has a tremendous impact on teaching many difficult engineering concepts. In fact the first textbooks using these tools to teach control system design have begun to appear [*Shahian and Hassul, 1993*]. Typical packages include EASY5x, MATRIXx, PSI/C, SIMNON, SIMULINK and VISSIM and a short description of each tool is given later. A more detailed comparison of EASY5x, MATRIXx and SIMULINK is also pursued.

Looking back across the three categories of dynamical systems and the tools which are available, these tools all use advanced graphics to establish the simulation diagram. For this reason, many products have been recently up-graded and reissued in new versions of the software. The addresses for obtaining more information on these simulation packages can be found at the end of the chapter. In the next section the general features of the graphical interface are discussed and the simulation tools ACSL (PROTOBLOCK), EASY5x, MATRIXx, PSI/C, SIMNON, SIMULINK and VISSIM are briefly described.

3.3 Dynamical System Simulation Tools : A Review

3.3.1 ICON Based Simulation

Advanced industrial system feasibility studies often require the development of a new set of system component simulation modules combined with a novel control system structure and integrated in a common software framework. The immediate problem for any project team is then to decide which software platform or package should be selected for the task. This used to be a fairly straight-forward decision since the existing in-house language, say FORTRAN, and its developed libraries would probably have been used. But now rapid changes in software products, in graphical interfaces and hardware configurations means that the decision has to be reviewed frequently and more carefully to ensure the long term viability of project work. Managers cannot afford to have the results of a new generation of development projects archived in nearly obsolescent information technology or jeopardised by a surprise future incompatible software release.

The control systems engineer usually needs two types of computer based tools to ease research and development work. One type provides comprehensive system simulation facilities and the other, control design routines. The latter have received extensive development and products like the MATLAB toolboxes or PROGRAM CC [*Johnson and Grimble, 1992*] are well established. System simulation facilities have been slower to develop and evolve, primarily because the tasks to be accomplished are not so readily formalised as neat self-contained algorithms.

The last decade has seen extensive developments in computer graphics and VDU technology. This has had an impact on system simulation tools enabling the symbolism inherent in system descriptions, namely the block diagram and its interconnections, to appear directly on the screen. The subroutine implementing the behaviour of the system component is buried in the simulation programme of the package. The fact that a graphical icon could represent a common system component meant that the interactive/pro-forma input of system model data was now possible in a transparent manner. If the system component is non-standard, then the creation of a customised icon is usually straight-forward.

In this way, the screen graphics of the new simulation tools mimic the system diagrams the modeller would sketch out prior to generating subroutine code. These packages collect together the various computer tools and component blocks, like simple integrators (with a choice of integration routines) and connection mechanisms, to form a toolbox for system modellers. The use of a Graphical User Interface (GUI) makes all the tasks of setting up a complex, possibly hierarchical, system simulation so much quicker, easier and efficient. These new icon based simulation tools have the following basic features in common:

- *Representation of mathematical models within the simulation tool environment.*

  A GUI facilitates the selection and interconnection of block icons by the user, and displays the model schematic.
- *Simulation.*

  The dynamic response of the system model to selected inputs is calculated using a numerical integration routine.
- *Graphics capability.*

  Data records from simulation are analysed and displayed using a graphics program.

- *Error Detection and Debug facility.*

  Errors in model definition or during simulation are indicated via messages. In addition there may exist a program or programs designed to detect the source of errors and possibly even explain why an error occurred, easing the debugging task.

- *User Communications.*

  The way in which the user communicates with the simulation tool varies from package to package. The two basic methods prevalent are instructions typed upon receiving acommand line prompt, or mouse operations. Modern simulation tools provide an on-line interactive environment. For example, the facility to pause a simulation, or to vary model parameters during simulation, or to be able to generate graphs as the simulation proceeds.

An important and continuing trend in the development of simulation tools is the progressive freeing of the user from low level coding tasks, allowing development of models at a higher conceptual level. At the level of individual blocks, libraries of standard components or industry specific process modules form the basic knowledge-base. Some industrial sectors, like chemical and process engineering are well advanced in this respect. Once the appropriate library is available, the graphical hierarchical structure available in the simulation tool allows the developer to concentrate on global system architecture and its dynamics. The advantage of the general purpose simulation tools is that they permit the creation of user specific modules across a wide range of technical areas, both small and large-scale. These tools are given a more detailed review in the next section.

3.3.2 Simulation Tools : A Short Survey

In order to select a simulation software for use, it is valuable to preview a number of packages. However, acquiring information about several tools can be a little hit and miss since the main sources of information are the professional and trade journals. One useful source of organised information can be obtained from the Working Group on Software (WGS), Eindhoven University of Technology in Holland. They have a database of system simulation and control design tools. This is an international group who release a Newsletter and database update. The WGS database currently has 26 entries describing system simulation software. The twenty six entries of the WGS database are far too many to review in detail. Those described here have been selected on the basis of visibility of use within the control professional community. A short description is given for each package and further detailed information can be obtained from the contact list given at the close of the chapter.

ACSL : The Advance Continuous Simulation Language (ACSL) has a structure based on Fortran which is specifically designed to ease the task of the system modeller and simulation development. This has been released in a Windows version giving such features as a project organiser which allows the efficient maintenance of system models, dialog box utility, a runtime window and a Help window. Important in this context are the graphical front-ends to ACSL; two are available: PROTOBLOCK and Graphic Modeller. These have all the hierarchical block build features expected of state-of the-art icon simulation tools.

EASY5x : EASY5x is an X-window version of Boeing's EASY5 Engineering Analysis System. EASY5 was introduced in 1987 and EASY5x released in 1992. EASY5x is used to model, simulate , analyse and design linear and nonlinear dynamic and static systems. It is capable of modelling large, complex, continuous, multi-rate digital, or hybrid systems.

The user interacts with the EASY5x program via a Motif-based graphical interface. The models can be constructed using blocks selected from EASY5x's standard components library or specialised libraries. The selection is made from pop-up menus, the required block position on the model schematic is indicated by pointing and clicking using the mouse. Interconnections between components are created by pointing and clicking on the 'from' and the 'to' blocks.

EASY5x facilitates hierarchical system representation to simplify complex model schematics. Again sub-models are expanded or reduced via a simple point and click using the mouse. Additional analysis tools form an integral part of EASY5x. These include a robust steady state finder, parameter optimisation, linear analysis and a model linearization facility. Results may be plotted on-line interactively or off-line. The plotting program presents a menu of command alternatives to the user below the graph view area, facilitating the quick manipulation of graphs.

MATRIXx : The structure of the MATRIXx environment has a great similarity to that of MATLAB. Many commands are shared in both, with minor syntax changes. Thus, MATRIXx is an interactive programmable matrix calculator with graphics capabilities. It was introduced in 1983 and provides solutions to complex, large scale matrix problems that arise in many engineering disciplines. It was introduced in 1983 and is now up to Version 3.

Built upon the MATRIXx computational environment is SYSTEMBUILD, facilitating graphical modelling and nonlinear system simulation. SYSTEMBUILD was first introduced in 1984. The SYSTEMBUILD editor utilises a mouse driven interface with pull-down menus and pop-up forms for entering model information.

Continuous, discrete, hybrid, event-driven and multi-rate digital systems can be simulated. Hierarchical systems can also be easily represented. Animated interactive simulation is also possible, with basic animated manual control panels that can be manipulated during the simulation. A library of common devices such as meters, gauges, and strip charts are provided to create a graphics display that is connected to the model, for monitoring. Model inputs and parameters may be altered via animated switches, dials and sliders whilst the simulation is running. The MATRIXx environment also provides a wide range of system design and analysis modules or toolboxes. These modules include control design, model reduction and optimisation amongst others.

Also XMATH has recently become available as a member of the MATRIXx product family. XMATH is an interactive mathematical analysis environment for X-window workstations, combined with a programmable graphical user interface. It combines a powerful object-oriented scripting language, 2D and 3D colour graphics and a library of numerical routines (including signal processing and control design functions).

PSI/C : This is a new C-version of the simulation package PSI originally produced at Delft Technical University, The Netherlands. It has been upgraded for interactive operation between the mouse and PSI menus. The graphics abilities have been enhanced. It is not an icon based simulation software as such since models are entered in PSI/C language with comments. It also has the facility to include new user written C-programs. This simulation package has found exposure in industrial and educational applications and is moderately priced.

SIMNON : This was developed at the Lund Institute of Technology, Sweden and been used there since 1974. SIMNON has been extensively revamped and extended. The basic software comes in five versions suitable for workstations and pc's. Most important is a new graphical user interface ISEE-SIMNON where complex hierarchical models may be constructed and investigated in a windows environment.

SIMULINK (MATLAB) : MATLAB is an interactive program designed to perform scientific and engineering numerical calculations. The program's fundamental data element is a matrix of arbitrary dimension. MATLAB was first introduced in 1981. Using the MATLAB numeric computation system is the SIMULINK program, designed for dynamic system simulation. SIMULINK became available in 1992.

SIMULINK facilitates the modelling and analysis of dynamic systems, including linear, nonlinear, continuous, discrete and hybrid model types. Models are constructed simply by taking components from a block library to a model worksheet and using the mouse to connect them up. SIMULINK can model systems in a hierarchical fashion by grouping portions of models into subsystems. Large and complex systems may be modelled.

Two important analysis tools are included in SIMULINK, model linearization and steady state operating point calculation. In addition the usual control design and analysis toolboxes are accessible from within the MATLAB environment. Such toolboxes include those for control design, μ-analysis and synthesis, neural networks and so on. MATLAB, together with SIMULINK and the above toolboxes forms an integrated environment for system design, modelling, simulation and analysis.

VISSIM : This is a new windows tool which uses a block diagram library to provide multi-rate continuous system simulation. There are 75 predefined blocks of both nonlinear and linear components which may be simulated using a choice of seven integration routines.

Animated graphics and graphics windows are also available. A compound block structure is obtainable for more complex multilevel systems.

It is of considerable interest to the new user to be able to compare simulation software, and establish a basis for comparison. Price is an obvious limiting factor in many applications, but since there is some evidence of widespread standardization of a software in some forward looking industrial and academic departments, a technical basis for comparison is of more than passing interest. In the final section of this chapter, three software tools are compared through a simple bench mark case-study.

3.4 An Assessment Exercise

3.4.1 The Assessment Basis

The three simulation packages chosen for assessment were EASY5x, MATRIXx and SIMULINK. The objective of the excercise was to compare and contrast the features and utilities of these tools for use on an industrial controls project. The long term objective is to create a library of process sub-system components, boilers, pumps, turbines, generators, valves and be able to connect them at will. In a layer above the unit layer the supervisory control behaviour is to be investigated, in particular, issues such as setpoint calculation, optimization and scheduling. The ability to examine global dynamic system behaviour simply and easily has an exciting potential for material and process energy savings. These types of objectives for an industrial simulation development should be fairly common. What then are the requirements sought for the simulation tool to be used?

In the following, the required attributes of the simulator are divided into those that are essential, and those that are desirable.

3.4.1.1 Essential Requirements

a) Capable of simulating complex systems which are:

- large scale (consisting of many interacting components)

- multivariable (many inputs, many outputs)
- linear and nonlinear
- mixed continuous + discrete system elements
- multirate in discrete time

b) Block diagrams including hierarchical system representation.

The ability to nest sub-units or sub-systems is an important facility. This enables the system to be constructed and analysed in a hierarchical manner. Thus the performance and behaviour at different system levels can be examined.

c) Equation based modelling.

When a new process unit model is developed, then setting up module code should be easy and closely related to the mathematical equation format.

d) High level code insertion facilities.

The ability to insert code from other sources (for example, other researchers, or even companies) means that the performance of different mixes of components can be investigated.

e) Interactive simulation capabilities.

What is meant here is the facility to perform *what if* analysis and also some provision for the generation of plots as the simulation proceeds.

f) Graphics capability.

Since the main output from the simulator will be graphical, the maximum flexibility in manipulating graphical output (scaling, text, labels, editing) is essential.

g) Fast runtime.

The automatic generation of runtime code from a model description, that is optimised in terms of its speed of execution, is a standard feature of current day simulation tools like ACSL. This is an essential requirement for efficient usage.

h) User friendly:

- well documented manuals
- easy to access/edit package functions
- error detection and debugging facilities
- good access to program suppliers

i) Reasonable costs

3.4.1.2 Desirable Requirements

a) Availability of related design tools.

For control systems development, these might typically include:

- matrix computation facilities
- time/frequency response tests and plots
- stability and sensitivity analysis
- equilibrium point determination
- signal processing routines
- model identification, transformation, reduction and linearization
- robust and optimal control

b) Access to a library of appropriate process or component modules.

The increasingly widespread availability of these icon based simulation tools is creating the opportunity to develop industry-based module libraries. It is always useful to try to identify the existence of such software. Such a library may save time and undue expense.

c) On-line help facility.

Access to a help index at the computer terminal, provision of hints, warnings, error messages is invaluable.

d) Workstation/PC compatibility.

The capability to operate on both workstation and PC computer platforms, with provision for easy transfer of models and data between platforms is very desirable.

e) Customised icon creator/editor.

Clearly a desirable feature to be able to generate industry specific symbols, for example a generator, or a distillation column icon.

f) Animated interactive simulation capability.

This allows for the inclusion of animated manual control panels that can be manipulated during the simulation. Model inputs and parameters may be altered via animated switches, dials, sliders etc. whilst the simulation is running. Thus for example, a simulated aircraft can be flown with animated controls whilst height, orientation, velocity and acceleration are all monitored via the animated instruments.

g) Optimised high level compiled code.

Facility to convert from a block diagram representation to speed optimised high level language source code, suitable for compilation. In this way the future software products developed can be protected from plagiarism by supplying the end-user with executable code only. Furthermore runtime would be reduced for a compiled model, compared to an interpreted model.

h) Automatic generation of real-time code.

This would facilitate hardware-in-the-loop testing of control designs generated from the research and development work, leading in turn to a functional prototype. The potential savings in terms of reduced programming effort required to produce real-time code are considerable. Consequently the commercial exploitation of the developed software products, would be eased.

With the essential and desirable attributes established, the three packages were used to simulate a simple boiler model. The attributes above were used to form a tabular comparison after the excercise was completed.

3.4.2 A Simple Boiler Model Test Case

Fig. 3.1 depicts a simple non-linear model of a drum-type boiler, representing mass flow and pressure effects [*de Mello, 1991*]. This model was derived from a more accurate model based upon fundamental energy, momentum and mass balances. It was demonstrated that under closed loop pressure control the simple model closely approximates the rigorous model's dynamic behaviour.

The main assumptions and features of the model along with some comments on these follow:

- The heat release in the furnace as a function of the fuel and air input signal is determined by the fuel system dynamic lags and dead-times associated with the coal feeder and pulveriser. By way of contrast, oil and gas fired furnaces exhibit faster response with a dynamic lag due to the fuel valve and air damper operation.

- The steam raised in the water-walls follows the furnace heat release with a dynamic lag due to the metal water-wall interface.
- The drum superheater pressure increases in proportion to the integral of the net steam flow into the lumped drum/superheater storage volume.
- The mass flow of steam from the drum to the superheater is proportional to the square root of the pressure drop between the drum and the superheater.
- The steam mass flow to the High Pressure turbine stage is regulated by the governor valve. The valve is modelled as a static device since its dynamics are relatively fast compared to those represented elsewhere in the boiler model.

The equivalent equation based state-space description of the simple boiler model block diagram was extracted from Fig. 3.1 and is presented below. These are the first steps an engineer would follow, namely the extraction of structure, equations and data of the model prior to using the simulation package.

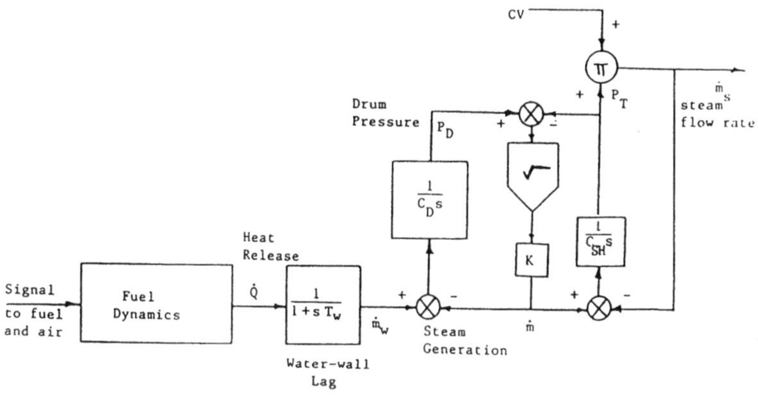

Fig. 3.1 A simple nonlinear model of a drum type boiler.

3.4.2.1 Dynamic State Space Boiler Model

INPUTS :

SIG	- Signal to fuel and air regulators
CV	- Governor Valve flow area

PARAMETERS:

T_W	- Water wall time constant
C_D	- Drum storage capacitance
C_{SH}	- Superheater storage capacitance
K	- Friction drop coefficient
τ_f	- Fuel system dead time
T_f	- Fuel system time constant

STATES:

$\dot{Q}$	- Furnace heat release
$\dot{m}_w$	- Mass flow rate of steam generated in the waterwalls (riser) entering the drum
P_T	- Superheater steam pressure at governor valve
P_D	- Saturated steam pressure in the drum

INTERNAL VARIABLES :

m	- mass flow rate of steam from drum to superheater

ALGEBRAIC EQUATIONS:

$$\dot{m} = k\sqrt{P_D - P_T}$$

$$\dot{m}_S = CV * P_T$$

DIFFERENTIAL EQUATIONS:

$$\frac{d}{dt}\dot{Q} = \frac{1}{T_f}\left[SIG(t-\tau_f) - \dot{Q}\right]$$

$$\frac{d}{dt}\dot{m}_w = \frac{1}{T_w}\left[\dot{Q} - \dot{m}_w\right]$$

$$\frac{d}{dt}P_T = \frac{1}{C_{SH}}\left[\dot{m} - \dot{m}_s\right]$$

$$\frac{d}{dt}P_D = \frac{1}{C_D}\left[\dot{m}_w - \dot{m}\right]$$

OUTPUTS :

$\dot{m}_S$ - superheated steam mass flow rate to high pressure turbine stage.

From the above, by setting the state time derivatives to zero it is straightforward to obtain the conditions for steady state equilibruim as:

STEADY STATE BOILER MODEL

$$\dot{Q} = \dot{m}_W = \dot{m} = \dot{m}_S = SIG$$

$$P_T = SIG/CV$$

$$P_D = (SIG/K)^2 + SIG/CV$$

These three equations provide the states and output that correspond to any given input values to give a steady state, or in other words a completely flat time response.

3.4.2.2 The Proposed Simulation Exercise

The model was initialised to the full load steady state operating condition. This corresponds to both inputs being set to unity. Note that the model employs the per-unit dimensioning system where the model pressures and mass flows are normalised by dividing by the full load rated drum pressure and output steam mass flow respectively.

This model was considered to be useful for comparing the simulation tools since:

- the model as presented in block diagram form should be quick to implement within the simulation tool environments.
- the model contains nonlinear elements (pure time delay, square root and a product), feedback loops and steady-state initialisation requirements, providing a nontrivial test case.
- the model represents a self-contained module as might be required for a power plant process library.

The objective of the assessment included items like ease of use, and the aesthetics of representation as well as the more formal items listed in section 3.4.1. The following program versions were used for the comparison:

EASY5x	Version 2.0
SIMULINK (MATLAB)	Version 1.0
SYSTEMBUILD (MATRIXx)	Version 3.0

3.4.3 The Assessment Results

3.4.3.1 The Block Diagram Schematics

The construction of the boiler model block diagram within each simulation environment proved to be a fairly straightforward task in each case.

Figures 3.2, 3.3, 3.4 depict the boiler model block diagram representation within SIMULINK, EASY5x and MATRIXx respectively. In the discussion that follows it will be seen that each package has relative advantages and disadvantages. Note that each package is capable of representing hierarchical block diagrams although this particular boiler model test case does not include a hierarchy.

SIMULINK's block diagram (Fig. 3.2) is relatively cluttered due to the inclusion of workspace blocks (required for recording simulation variables data), scopes (required for interactive graphics) and a clock (required for recording simulation time data). Also in SIMULINK the integrator block has a fixed gain of unity, therefore an additional gain block is required with each non-unity gain integrator.

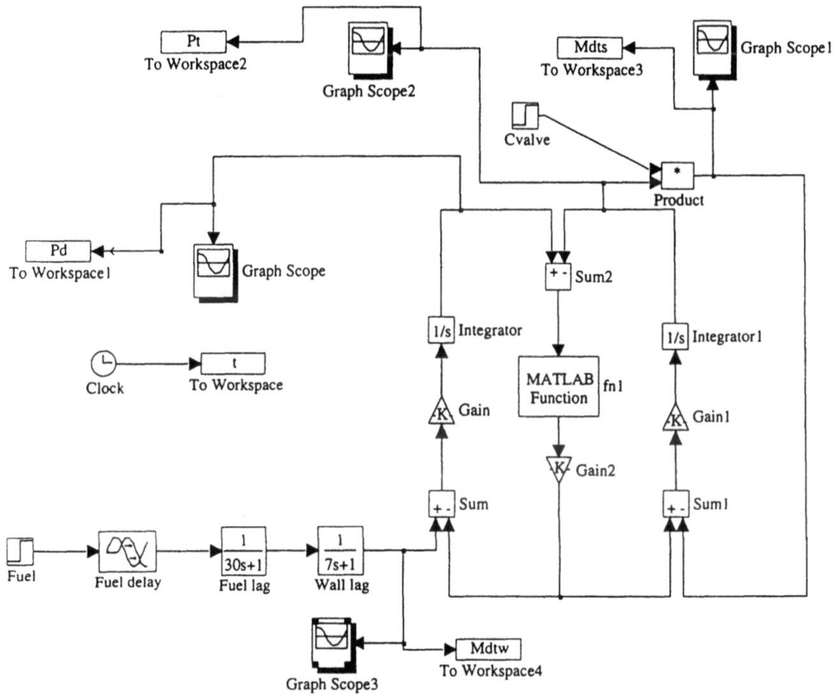

Fig. 3.2 SIMULINK - boiler model block diagram representation.

By comparison EASY5x includes all of the above information in lookup tables external to the block diagram schematic, making for a cleaner schematic (Fig. 3.3). Furthermore EASY5x avoids the use of step function blocks (although this can be achieved in the other packages as well). It would appear to be better to show a step input on the schematic, thus giving a clear indication of the nature of the input. Indeed the EASY5x block diagram could be simplified even more by the removal of the input gain blocks. However these were included to clearly show the location of the inputs.

In terms of visual appeal, EASY5x arguably has the most impressive block diagram schematics. Only EASY5x has an icon editor for the creation of customised icons.

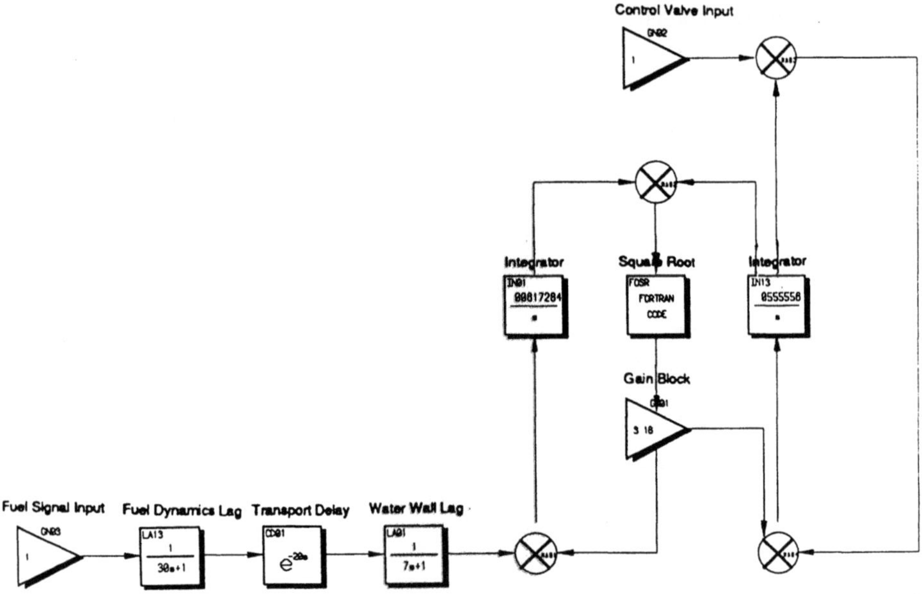

Fig. 3.3 EASY5_x - boiler model block diagram representation.

The MATRIXx block diagram (Fig. 3.4) is somewhat intermediate in complexity. Step inputs are clearly shown, however the step block has a fixed initial output condition of 0, necessitating the inclusion of a constant block and summer block to simulate a step from 1 to 1.02. This is avoided in SIMULINK.

The labelled output tags in the MATRIXx schematic are equivalent to the workspace blocks in SIMULINK. MATRIXx provides the neater solution in this case.

Manipulation of the block diagram schematic to eliminate connection crossovers required the greatest effort in MATRIXx. Fig. 3.4 shows that one crossover remains. There is a way to remove this crossover, but the point is that these are relatively difficult to avoid in MATRIXx. Each component has a choice of icons available to represent it. On a trial and error basis the icons are sorted through (including the 'flipped' versions) until no connection crossovers occur. It is observed that for SISO blocks, MATRIXx only allows the input and output to lie horizontally. EASY5x and SIMULINK include vertical orientation as well. This may be a contributory factor in producing the crossover hazard which sometimes occurs in MATRIXx.

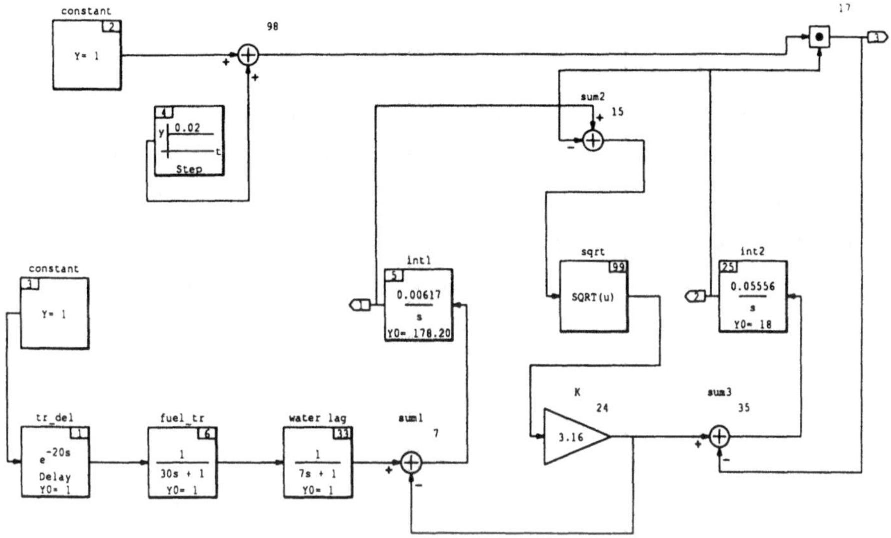

Fig. 3.4 $MATRIX_x$ - boiler model block diagram representation.

Only in SIMULINK is the routing of connections manual, EASY5x and MATRIXx employ automatic router programs (in MATRIXx it is possible to manually adapt routes after automatic routing has completed, with certain limitations).

In conclusion the main differences existing between the schematics fall into two main areas:

- the amount and type of information displayed.
- what information can be manipulated by the user and what information is fixed by the simulator.

3.4.3.2 Model Initialisation

It is often necessary to initialise a dynamic model at a certain steady-state operating condition. This results in a completely flat time response provided the inputs are not altered from their initial values. Thus the problem of having to wait for initial transients to die out before applying a test input signal (e.g. a step change) is removed. The initial transient may cause the simulation to become unstable and/or crash if the model is driven outside of its operating envelope. If these do not occur there is still the potential problem of the initial transient taking a long time to decay.

In order to initialise the model to steady-state, the internal model states and model outputs must be set to values that correspond to the steady-state condition for the given inputs. For simple models, by considering the algebraic relationships inherent in the model, it is possible to construct a set of equations that prescribe the steady-state values of the model states and outputs as a function of the steady-state input values. For more complex models, this task becomes formidable. However many simulation tools include a program that automatically locates steady-state operating points or 'equilibrium points'. This is achieved without having to perform any simulation.

Prior to dynamic simulation, the boiler model was required to be initialised at full load steady state operation. The initial values of the model variables were calculated using the steady state equations given above. These values were then compared to the results of initialization procedures within the various software tools.

In MATRIXx it was easy to specify initial conditions on each dynamic block, therefore the model initialisation task was performed quickly and without problem.

In EASY5x, the continuous transport delay block did not include the facility to initialise its output value. Neither does the reference manual indicate what determines the initial output value. In fact, the initial value of the input is assigned to the output. Because of this, in order for the block to work it is necessary for a dynamic block with a specifiable initial output value to precede it. This is why the ordering of the fuel dynamics elements is reversed in the EASY5x schematic compared to SIMULINK's and MATRIXx's. Furthermore extra table sizing information must be provided in order for the block to work properly, making its implementation relatively complicated compared to its equivalents in SIMULINK and MATRIXx.

For SIMULINK, unfortunately the 1st order lag blocks (or any other transfer function block) do not have an explicit means of initialising their outputs to a value other than zero. The reference manual describes a method where individual block outputs can be initialised from the command line. However this requires extra effort from the user, involving consideration of the ordering of dynamic block states after transformation of the block transfer function into state-space companion form. This could become very tedious for large scale models.

3.4.3.3 Dynamic Simulation

The time responses to a 2% step increase in each model input were simulated. The simulation time increment was selected as 1 second, and the simulation was performed over 1000 seconds.

All three simulation tools provide a choice of integration algorithms. It was found that a number of algorithms provided satisfactory performance. Good quality results with relatively short runtimes were obtained using the following options in each package:

EASY5x	- BCS Gear algorithm (the default option)
SIMULINK	- Gear algorithm
MATRIXx	- Variable-step Kutta-Merson algorithm (the default option)

It was not possible to make a meaningful comparison of runtime speed between the simulation packages. This was due to the small-scale of the test case. In addition, it would be necessary to have an isolated host computer, to ensure equality in hardware terms. At the Industrial Control Centre's installation, SIMULINK and EASY5x are node locked to separate workstations existing within a work group, whereas MATRIXx is installed to 'float' upon the same work group. For the simplified boiler model test case each simulation was completed within 10 seconds.

3.4.3.4 Graphical Output

Figures 3.5, 3.6, 3.7a, and 3.7b depict one particular step response for each simulation package. It is noted that the responses are the same in each package, also they agree with the original results of de Mello *(1991)*. The final steady state values are those predicted by solving the steady state equations given above.

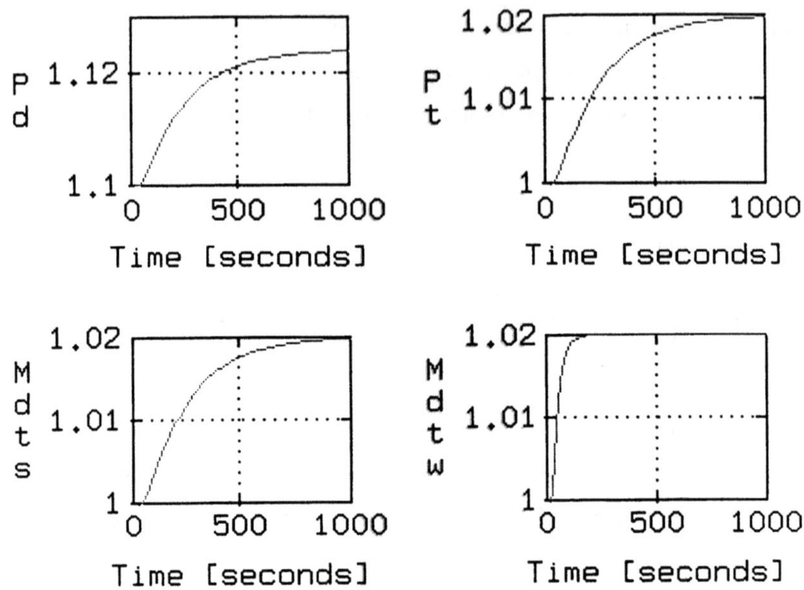

Fig. 3.5 SIMULINK - plot of responses to a 2% fuel step input.

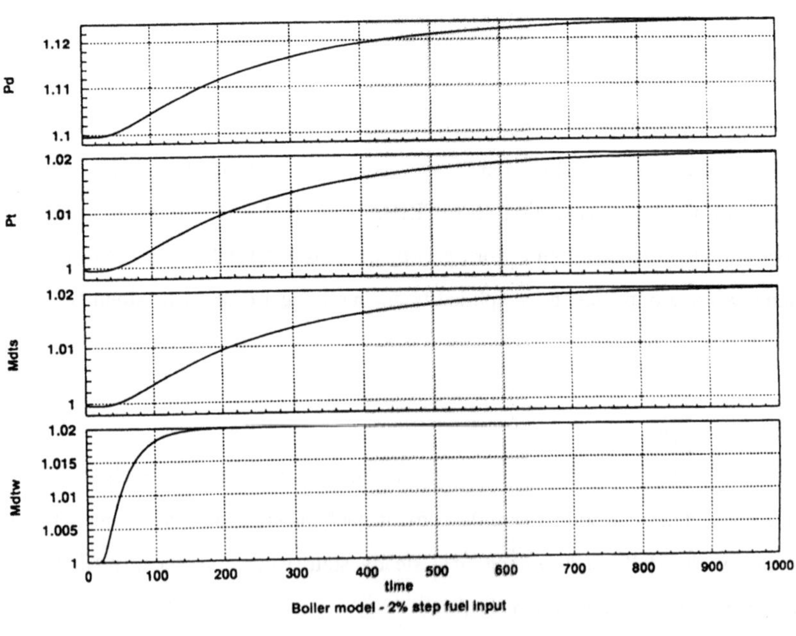

Fig. 3.6 MATRIXx - plot of responses to a 2% fuel step input.

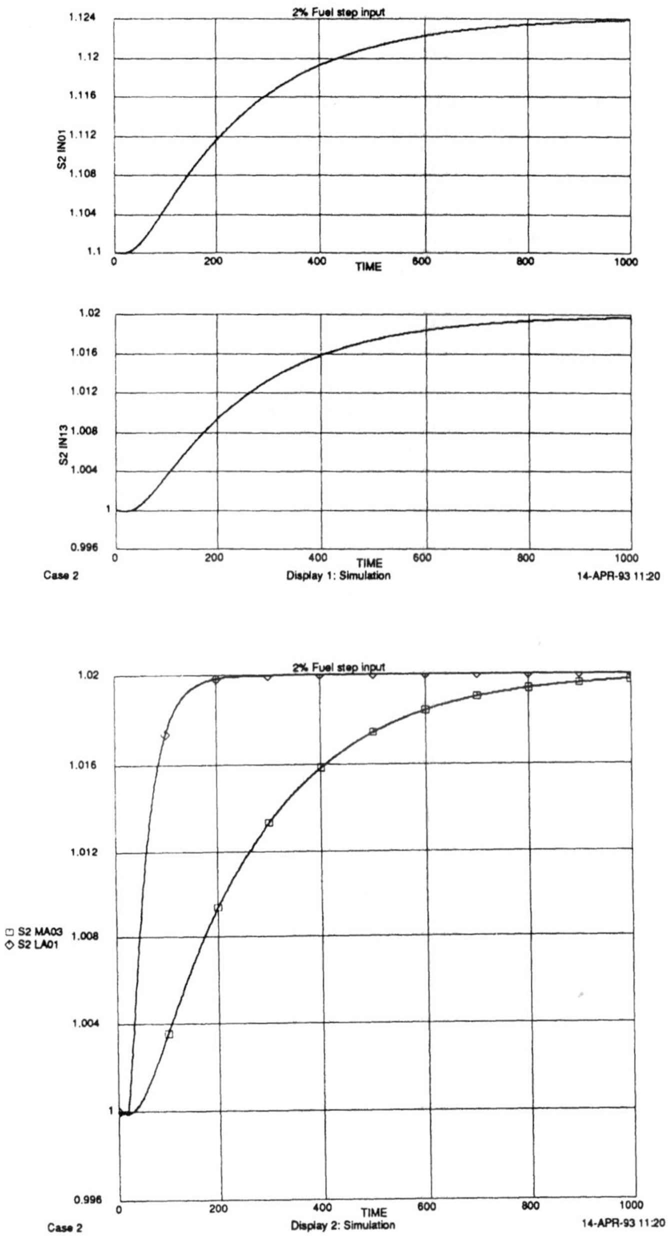

Fig. 3.7 EASY5x - plot of responses to a 2% fuel step input.

The SIMULINK graph (Fig. 3.5) was obtained by use of the SIMULINK 'subplot' command, enabling the screen to be divided into four views.

Similarly the MATRIXx graph (Fig. 3.6) was obtained by use of the MATRIXx 'strip' plot command. The MATRIXx graphs are neater than SIMULINK's in this particular case.

For the EASY5x graphs (Figs. 3.7a and 3.7b), it was more convenient - due to inherent scaling restrictions when graphs are grouped into one display - to create separate figures. Unfortunately it was not possible to alter the displayed model variable names using the EASY5x plot program. For example the user is stuck with 'S2 IN01' rather than 'Pd' to represent drum pressure.

SIMULINK has the disadvantage that the graphics program is not menu driven.

EASY5x graphics program provides a Motif-based user interface, a menu of command options is located below the graph view area. This was found to be particularly convenient when manipulating graphs.

The MATRIXx environment includes the XMATH tool. This also includes a Motif-based user interface for generating graphs. Thus the XMATH graphics offers a valuable alternative to those of MATRIXx.

Finally both SIMULINK and MATRIXx environments facilitate the generation of 3D mesh graphs whereas EASY5x is restricted to 2D graphs.

3.4.3.5 Summary Table

The table shows a quick summary of the features available and performance relating to the three simulation tools considered. The areas assessed correspond to the essential and desirable simulation tool features listed above.

3.5 Conclusions

It is a measure of computing technology progress that a decade ago, the emphasis of a simulation package assessment would have concentrated on speed, memory, and access to numerical algorithm libraries. These new simulation tools are quick and efficient. Our experience has been that many industrial users see these as the new media for developing and archiving system simulations. The above exercise showed that all three packages have the same basic facilities and that their main differences lie in additional features of the package, the style of screen representation and appearance and the capabilities of the graphics plotting facilities. Finally, there may be some very precise requirement of a particular project which has a decisive influence on final choice. However, there is no more useful exercise than several free trials which are usually very easy to arrange.

FEATURE	**EASY5x**	**SIMULINK family**	**MATRIXx family**
Essential Features			
a) Capable of simulating complex systems	X	X	X
b) Block diagram facility	X	X	X
c) Equation based models	X	X	X
d) Code insertion into block diagram	X	X	X
e) Interactive simulation	X	X	X
f) Graphics program	X	X	X
g) Code run-time	NA	NA	NA
h) User-friendly	X	X	X
Desirable Features			
a) Design tools	XL	X	X
b) System libraries	X		
c) On-line help	X	X	X

d) Operation on workstation and PC	-	X	X
e) Customised icons	X	-	-
f) Animated interactive simulation	-	-	X
g) Blk. diagm to high level source code conversion	X	-	X
h) Real time capability	X	-	X

Table 3.1 Simulation tools features. Key : X = *Feature available*, XL = *Feature available but relatively limited*, - = *Feature not available*, NA = *Not assessed.*

3.6 References for the Chapter

Ho Y. ***Dynamics of discrete event systems,*** Proc. IEEE, Vol. 77, No.1, 1989.

Johnson M.A. and Grimble M.J ***Using Program CC for Control System Design***, Measurement and Control, Vol. 25, pp 144-148, June., 1992.

de Mello F.P. ***Boiler Models For System Dynamic Performance Studies***, IEEE Trans. on Power Systems, Vol. 6, No. 1, pp 66-74, Feb. , 1991

Ramadge P. and M. Wonham ***Supervisory control of a class of discrete event processes***, SIAM Journal of Control and Optimization, Vol. 25, 206-230, 1987.

Shahian B. and M. Hassul ***Control System Design using MATRIXx***, Prentice Hall Inc., Englewood Cliffs, USA, 1993

3.7 Software Suppliers

- SIMAN/CINEMA:

  The Cimulation Centre Ltd, Cornbrash Park, Bumpers Way, Chippenham, Wilts, SN14 6RA, U.K.

- SIMPLE:

  AESOP Ltd, Preston Technology Management Centre, Marsh Lane, Preston, PR1 8UD, U.K.

- SIMFACTORY/SIMPROCESS:

  CACI Products Division, Coliseum Business Centre, Watchmoor Park, Riverside Way,Camberley,Surrey, GU15 3YL, U.K.

- WITNESS:

  AT and T ISTEL, Highfield House, Headless Cross Drive, Redditch, B97 5EQ, U.K.

- CADAS:

  SIMRAD ALBATROSS As, Po Box 483, N-3601, Kongsberg, NORWAY.

- ASPENPLUS, MAX, SPEEDUP:

  Aspentech U.K., Sheraton House, Castle Park, Cambridge, CB3 0AX, U.K.

- ACSL:

  Rapid DATA Ltd., Crescent House, Cresent Road, Worthing, BN11 5RW, U.K.

- EASY5x:

  Boeing PO Box 24346, MS 7L-46, Seattle, WA 98124-0346, U.S.A.

- MATRIXx:

  Integrated Systems Inc., 1st Floor, Gatehouse, Fretherne Road, Welwyn Garden City, Herts, AL8 6NS, U.K.

- PSI/C:

  BOZA Automatisering Bv, PO Box 264, 5670 AG Nuenen, THE NETHERLANDS.

- SIMNON:

  SSPA, Box 24001, S-40022 Goteborg, SWEDEN.

- SIMULINK:

  The MathWorks Inc., 24, Prime Park Way, Natick, MA 01760, U.S.A.

- VISSIM:

  Adept Scientific Micro Systems Ltd., 6 Business Centre West, Avenue One, Letchworth, Herts, SG6 2HB, U.K.

- For **Working Group on Software**, contact:

  R. Kool, Eindhoven University of Technology, Dept of Mathematics and Computing Science, PO Box 513, 5600 MB Eindhoven, THE NETHERLANDS.

CHAPTER 4

PROCESS MODELS

4.1 Introduction

This chapter presents mathematical models for the components of a typical CHP configuration. In this particular study the typical plant was taken to be that of Skegton unit, the structure of which was described in Chapter 1. The component models required are : boiler, gas turbine, steam turbine, condenser, electrical generator, gas merge and split, valve, pump, deaerator, actuator and pipe. In addition lower level component controllers for the boiler, gas turbine, condenser and the deaerator are modelled.

The complexity of the models was chosen to suit the objectives of the research into supervisory control strategies. This influenced among other things, the number of inputs and outputs represented in any particular model, and the required model accuracy. The models are based on first principles physical and thermodynamic laws with appropriate simplifications. Sometimes heuristic relationships are used. The models are often strongly dependent on the existing achievements listed in the cited literature. The literature review sections of this chapter are not meant to be complete overviews of the history and current status of modelling power plant components. Their purpose is simply to highlight the different approaches applied in modelling a particular component, and to focus upon the most relevant techniques suitable for the purposes of this research. The reader is referred to the review papers of [*Maffezzoni, 1992*], [*IEEE, 1983, 1988*] for a fuller coverage of the published literature relating to power plant modelling.

For more complicated components like boiler, gas turbine, steam turbine and heat exchanger, the models are divided into several parts reflecting the actual physical construction of a component. The controllers modelled are based upon real designs as reported in the cited literature. The models are mostly dynamic, represented by nonlinear, continuous-time state-space equations.

The chapter opens with a brief description of state space modelling since all the component models are formulated in state space. The component models are then presented systematically.

4.2 State Space Modelling

A brief outline of state space modelling concepts is provided in this section. A mathematically convenient model structure for control design and analysis is known as the state space model. The general form of a state space model for a continuous system is given by;

$$\dot{x} = f\left(x(t), u(t), t\right) \quad , \ 0 \le t < \infty$$

$$y = g\left(x(t), u(t), t\right) \quad , \ 0 \le t < \infty$$

where :

t	time; scalar
x(t)	system state; *n* dimensional vector
u(t)	system input; *m* dimensional vector
y	system output; *r* dimensional vector
f	nonlinear function; $\Re^n \times \Re^m \times \Re_+ \rightarrow \Re^n$
g	nonlinear function; $\Re^n \times \Re^m \times \Re_+ \rightarrow \Re^r$

The *state* of a system is defined as being any property of the system such that knowledge of the input and state for $t \ge 0$ uniquely and completely determines the output for $t \ge 0$. The *state space* is the set of all states of a system.

Note that the state space representation of a system is not unique. The state variables may be selected to provide a convenient mathematical structure (as is the case for state canonical and companion forms) or they may be based upon physical variables existing in a system.

All the equation based models described in this chapter are in state space form, with the states chosen to represent physical variables (e.g. pressures, temperatures, flows etc.). Where a model is described by a simulation block diagram, as is the case for low level controllers , this is converted to state space form by MATRIXx prior to simulation. Thus a state space model is defined for the complete Skegton unit configuration.

There are some useful advantages of adopting the state-space modelling approach :

(i) The modelling approach based on *physical* equations often leads quite naturally to (nonlinear) state-space forms.
(ii) The state-space method is *systematic* and creates a precise framework for enumerating inputs, outputs, state variables and system parameters.
(iii) For the expert, there is an extensive class of useful algebraic and numerical analysis techniques available. There is also a well established literature.
(iv) The MATRIXx linearisation and steady-state location algorithms assume a state-space system description.

4.3 Skegton Unit Component Models

4.3.1 Boiler

4.3.1.1 Introduction

A typical drum-type boiler is depicted in Fig. 1.12 of Chapter 1. The feedwater is supplied to the drum. In the drum the evaporation of water occurs . The water flows into the downcomers, then enters the risers. In the risers the heat from the furnace is used to increase the water temperature and eventually to cause evaporation. Thus the circulation of water, steam, and water and steam mixture takes place in the drum, the downcomers and the risers.

Steam generated in the risers is separated in the drum from where it flows through the superheater on to the high pressure turbines. It may then be recycled to the boiler in the reheater where its energy content is increased.

Desuperheating spray water is introduced in the superheater for control of main steam temperature.

The combustion products path is also shown. The risers absorb radiant heat in the furnace. The hot gases leaving the furnace transfer heat by radiation and convection to the superheater. Then heat is transferred by convection to the reheater and finally to the economiser, before exiting the boiler via the stack. The burner tilt is used to change radiation heat distribution between the risers and the superheater.

This section describes a model of the boiler. The equations presented are mostly based on first principles thus are general enough to represent a wide range of boiler sizes (from 1 MW to 600 MW). The boiler is divided into six modules corresponding to its physical structure. The modules are : furnace, risers, downcomers, drum , superheater, reheater. The combustion equations are valid for gas, liquid and fossil fuels.

4.3.1.2 Literature Review

Many papers present models of boilers at different levels of complexity. Fairly complicated models may be found in [*Chien et al, 1958*], [*Cori and Busi, 1977*], [*McDonald and Kwatny, 1970*]. These models have more than ten nonlinear differential equations and over one hundred non-linear static equations describing the design of all parts of the boiler.

Another class of models are derived for very limited purposes e.g. overall performance of the power station or control of only one particular variable. In this case many dynamic equations are neglected and many variables are not included. Examples of this type of simple model may be found in [*Chiacciarini et al, 1991*], [*Astrom and Bell, 1988*], [*Cheres, 1990*], [*de Mello, 1991*] and [*Ray and Majumder, 1985*].

For the purpose of this work the model required is something between the two categories described above. It must be detailed enough to include the dynamics of all variables which are important in CC/CHP applications but, on the other hand, it must be simple enough to permit the construction of an effective control algorithm.The boiler model presented here is based mainly on the paper [*Chien et al, 1958*] with some extensions from the paper [*Nicholson, 1964*]. The furnace model was created from the papers [*Cori and Busi, 1977*], [*Nicholson, 1964*] and from the book [*Rhine and Tucker, 1991*]. The attemporation flow was incorporated into the model in a manner similar to that of [*Thomas et al, 1976*]

4.3.1.3 Model Assumptions

The main assumptions used to construct the model are as follows:

(i) The model only includes time derivatives of variables, spatial derivatives were not considered.

(ii) Polynomial fits to steam tables were used to establish the relations between steam parameters such as enthalpy, density, temperature, pressure.

(iii) Superheated steam and furnace exhaust gases were treated as ideal gases.

(iv) Constant volumetric flow maintained in downcomer by circulation pumps.

(v) Mass flow dynamics for riser and reheater modelled as a 1st order lag.

(vi) The model includes only the main parts of the boiler using lumped characteristics for those parts which consist of more than one section (e.g. two sections of superheater and an attemporator between them, these are lumped to form one section called the superheater plus attemporator).

(vii) That the variables in the model are assumed to satisfy basic physical thermodynamic balances as follows:

◊ ***Heat balance for a heat exchanger in the form:***

$$Q_{in} + w_{in} h_{in} = w_{ou} h_{ou} + V \frac{d}{dt}(\rho_{ou} h_{ou})$$

where :

Q_{in}	*incoming heat flow*
w_{in}	*inlet mass flow*
h_{in}	*inlet specific enthalpy*
w_{ou}	*outlet mass flow*
h_{ou}	*outlet specific enthalpy*
ρ	*specific density*
V	*volume*

◊ ***Mass balance in the form***:

$$w_{in} - w_{ou} = \frac{d}{dt}(\rho_{ou} V)$$

◊ ***Friction losses (momentum balance)***

$$p_{in} - p_{ou} = \gamma \frac{w_{in}^2}{\rho_{in}}$$

where :

p_{in} *inlet pressure*
p_{ou} *outlet pressure*
γ *friction coefficient*

(viii) The heat transfers due to radiation and convection are modelled using:

◊ ***Stefan - Boltzman law for radiation***

$$Q = K\theta w_g T_g^4 \frac{1}{\rho_g}$$

where :

Q *heat flow*
K *a coefficient*
θ *spheric angle*
w_g *flow of substances entering combustion*
T_g *temperature of combustion gases*
ρ_g *density of combustion gases*

◊ ***Experimental heat transfer equation combustion gas-metal*** *(assuming turbulent combustion gas flow):*

$$Q = Kw_g^{0.6}(T_g - T_m)$$

where :

T_g *gas temperature*
T_m *metal temperature*

◊ ***Experimental heat transfer equation metal-steam*** *(assuming turbulent steam flow):*

$$Q = Kw_s^{0.8}(T_m - T_s) \quad \text{or} \quad Q = K_r(T_m - T_s)^3$$

where : T_S *steam temperature*

4.3.1.4 Boiler Equations

The boiler is divided into six interconnected subsystems. They are:

- furnace
- riser
- downcomer
- drum
- superheater plus attemporator
- reheater (optional)

<u>The Furnace</u>

The furnace is described by the following equations:

Heat balance for combustion

$$C_F w_F + h_A w_A + h_G w_G - Q_{ir} - Q_{is} - w_{EG} R_s \left(1 + \frac{y}{100}\right) \cdot h_{EG} = V_F \frac{d}{dt}\left(\rho_{EG} h_{EG}\right)$$

where

C_F -	*fuel calorific value*
w_F -	*fuel flow*
h_A -	*air specific enthalpy*
w_A -	*air flow*
h_G -	*exhaust gas (from the gas turbine) specific enthalpy*
w_G -	*exhaust gas (from the gas turbine) flow*
Q_{ir} -	*heat transferred by radiation to risers*
Q_{is} -	*heat transferred by radiation to the superheater*
w_{EG} -	*gas mass flow through the boiler*
h_{EG} -	*gas specific enthalpy*
R_S -	*stoichiometric air/fuel volume ratio*
y -	*the percentage excess air level*
V_F -	*combustion chamber volume*
ρ_{EG} -	*gas density*

Mass balance for combustion

$$w_F + w_A + w_G - w_{EG} = V_F \frac{d}{dt}\rho_{EG}$$

Heat transferred by radiation to risers

$$Q_{ir} = \theta k V_F \sigma T_g^4 \frac{1}{\rho_{EG}}$$

where

T_g - *gas temperature*
σ - *Stefan-Boltzman constant*
k - *attenuation coefficient*
θ - *burner tilt angle*

Heat transferred by radiation to the superheater

$$Q_{is} = (1-\theta) K V_F \sigma T_g^4 \frac{1}{\rho_{EG}}$$

The total heat transferred to the superheater

$$Q_{gs} = Q_{is} + k_{gs} w_{EG}^{0.6}(T_{gs} - T_{st})$$

where

T_{gs} - *gas temperature at the superheater*
T_{st} - *superheater metal tube temperature*
k_{gs} - *an experimental coefficient*

Heat transferred by convection to the reheater

$$Q_{rs} = k_{rs} w_{EG}^{0.6}(T_{gr} - T_{rh})$$

where

T_{gr} - *combustion gas temperature at the reheater*
T_{rh} - *reheater metal tube temperature*
k_{rs} - *an experimental coefficient*

Heat transferred by convection to the economiser

$$Q_{es} = k_{es} w_{EG}^{0.6} (T_{ge} - T_{et})$$

where

T_{ge} - *combustion gas temperature at the economiser*

T_{et} - *economiser metal tube temperature*

k_{es} - *an experimental cofficient*

Heat balance for combustion gases flowing across the superheater surface

$$Q_{gs} = w_{EG} c_{gs} (T_g - T_{gr}) + Q_{is}$$

where c_{gs} - *combustion gas specific heat capacity*

Heat balance for combustion gases flowing across the reheater surface

$$Q_{rs} = w_{EG} c_{gs} (T_{gr} - T_{ge})$$

where T_{ge} - *gas temperature at the economiser*

Heat balance for combustion gases flowing across the economiser surface

$$Q_{es} = w_{EG} c_{gs} (T_{ge} - T_{g1})$$

where T_{g1} - *boiler exhaust gas temperature*

Percentage excess air

$$y = 100(w_A + \gamma_A w_G - w_F R_s) \frac{1}{w_F R_S}$$

where γ_A - *content of fresh air in exhaust from gas turbine*

Exhaust gas flow through the boiler

$$w_{EG} = k_f p_G$$

where

k_f - *a friction coefficient*

p_G - *gas pressure in furnace*

Furnace gas pressure

$$p_G = R_{EG}\,\rho_{EG}\,T_g$$

where R_{EG} - *ideal gas constant for exhaust gases*

Furnace gas temperature:

$$T_g = \frac{h_G - h_{ref}}{c_{pg}} + T_{ref}$$

where

h_{ref} - *reference exhaust gases enthalpy condition*

T_{ref} - *reference exhaust gases temperature condition*

c_{pg} - *specific heat of exhaust gases at constant pressure*

Input variables to the furnace:

w_F - *fuel flow*

w_A - *air flow*

h_G - *exhaust gas (from the gas turbine) specific enthalpy*

w_G - *exhaust gas (from the gas turbine) flow*

θ - *tilt angle*

Variables specified in other parts of the boiler:

T_{st} - *superheater metal tube temperature*

T_{rh} - *reheater metal tube temperature (from the reheater block)*

T_{et} - *economiser metal tube temperature (from the economiser block)*

Output variables from the furnace model, supplied to other parts of the boiler:

Q_{ir} - *heat transferred to risers,*

Q_{gs} - *heat transferred to the superheater*

Q_{rs} - *heat transferred to the reheater*

Q_{es} - *heat transferred to the economiser*

p_g - *furnace air pressure*

The Risers

The risers are described by the following equations:

Mass balance

$$w_d - w_r = V_r \frac{d}{dt}(\rho_r)$$

where

w_d - *water mass flow from the downcomer*

w_r - *riser liquid-vapour mixture mass flow*

V_r - *riser volume*

ρ_r - *liquid-vapour mixture density at the riser*

Heat balance

$$Q_r + w_d h_w = w_r h_r + V_r \frac{d}{dt}(\rho_r h_r)$$

where

Q_r - *heat transferred to steam*

h_w - *specific enthalpy of downcomer and drum water*

h_r - *specific enthalpy of liquid-vapour mixture*

Mixture equation

$$\rho_r = \left[\frac{x}{\rho_v} + \frac{(1-x)}{\rho_{wv}}\right]^{-1}$$

where

x - *steam quality*

ρ_v - *density of saturated steam*

ρ_{wv} - *density of saturated water*

Gas-tube heat balance

$$Q_{ir} = Q_r + M_r c_{rt} \frac{d}{dt}(T_{rt})$$

where

Q_{ir} - *heat transferred by radiation to risers (from the furnace model)*

Q_r - *heat transferred from the tube to steam*

M_r - *metal mass*

c_{rt} - *metal specific heat*

T_{rt} - *metal tube temperature*

Heat transferred to steam

$$Q_r = k_r(T_{rt} - T_v)^3$$

where

T_v - *riser steam temperature*

k_r - *an experimental coefficient*

***Flow relation** - downcomers to risers*

$$\dot{w}_r = (w_d - w_r)/\tau_r$$

where : τ_r - an empirical flow time constant

Variables incoming to the risers from other modules of the boiler :

w_d - *water mass flow from the downcomer*

Q_{ir} - *heat transferred to the risers (from the furnace)*

h_w - *specific enthalpy of downcomer and drum water (from the drum)*

h_v - *specific enthalpy of saturated steam (from the drum)*

h_{wv} - *specific enthalpy of saturated water (from the drum)*

T_v - *steam temperature (from the drum)*

p_w - *water drum pressure (from the drum)*

p_v - *steam drum pressure (from the drum)*

ρ_w - *water density (from the drum)*

Outputs from the risers, connected to other blocks of the boiler:

w_r - *liquid-vapour mixture mass flow (to the drum)*

x - *steam quality (to the drum)*

The Drum

The drum is described by the following equations:

Liquid mass balance

$$w_e + (1-x)w_r - w_d - w_{ec} = \frac{d}{dt}\left(m_{dl}\right)$$

where

w_e - *feedwater flow (from the economiser)*

x - *steam quality (from the risers)*

w_r - *liquid-vapour mixture mass flow (from the risers)*

w_d - *water mass flow out to the downcomer*

w_{ec} - *drum liquid mass evaporation*

m_{dl} - *drum liquid mass*

Drum liquid level

$$\frac{m_{dl}}{\rho_w} = f(L) \qquad \text{thus} \qquad L = f^{-1}\left(\frac{m_{dl}}{\rho_w}\right)$$

where :

ρ_w - *drum water density*

L - *drum water level*

f - *a function depending on the shape of the drum*

For example, for an upright cylinder the equation is:

$$V_L = f(L) = \pi r^2 L$$

for a sphere the equation is:

$$V_L = \frac{1}{3}\pi L^2 (3r - L)$$

A typical industrial drum shape is that of a cylinder with hemispherical ends laid horizontally. For this shape it can be shown that

$$V_L = \tfrac{1}{3}\pi L^2 (3r - L) + \tfrac{1}{2}(W - 2r) r^2 (\theta - \sin\theta)$$

where $\theta = 2\cos^{-1}\left(\frac{r-L}{r}\right)$

where

r - cylinder radius

W - cylinder length

Downcomers mass flow

$$w_d = v_{dow}\, \rho_w$$

where v_{dow} = volumetric liquid flow rate through downcomer.

Liquid heat balance

$$w_e h_e + (1-x) w_r h_{wv} = w_d h_w - w_{ec} h_v + \frac{d}{dt}\left(m_{dl} h_w\right)$$

where :

h_e - *feedwater specific enthalpy (from the economiser)*

h_{wv} - *specific enthalpy of saturated water*

h_w - *specific enthalpy of drum water (going to the downcomer)*

h_v - *specific enthalpy of saturated steam*

Steam mass balance

$$w_{ec} - x w_r = w_v + \frac{d}{dt}\left(V_v \rho_v\right)$$

where :

w_v - *steam mass flow out of the drum (to the superheater)*

V_v - *vapour volume*

ρ_v - *steam density*

Evaporation dynamics

$$w_{ec} = K_{ec}\left(T_w - T_v\right)$$

where :

K_{ec} - *a coefficient*

T_w - *water temperature in the drum*

T_v - *saturated steam temperature in the drum*

Vapour volume

$$V_v = V - V_L$$

where :

V - *drum volume*

V_L - *drum liquid volume*

***Variables incoming from other modules of the boiler*:**

x - *steam quality (from the risers)*

w_r - *liquid-vapour mixture mass flow (from the risers)*

w_v - *steam mass flow out of the drum (from the superheater model)*

w_e - *feedwater flow (from the economiser)*

h_e - *feedwater specific enthalpy (from the economiser)*

***Outputs from the drum model, connected to other modules of the boiler*:**

w_d - *water mass flow out of the drum (to the risers)*

p_v - *steam drum pressure (to the risers and to the downcomer)*

p_w - *water drum pressure (to the risers and to the downcomer)*

ρ_w - *water density (to the risers and to the downcomer)*

ρ_v - *saturated steam density*

h_v - *saturated steam specific enthalpy (to the riser)*

h_w - *drum water specific enthalpy (to the riser)*

h_{wv} - *specific enthalpy of saturated water (to the riser)*

T_v - *temperature of saturated steam (to the riser)*

The above variables (except w_d) are calculated from saturated steam and saturated water tables given ρ_v - *density of saturated steam* and h_w - *specific enthalpy of drum water.*

Output from the drum model which is also an output from the boiler model :

L - *drum water level*

Superheater plus attemperator

The superheater plus attemperator equations are:

Mass balance

$$w_v - w_s + w_a = V_s \frac{d}{dt}(\rho_s)$$

where :

w_v - *steam mass flow from the drum to the superheater*

w_s - *steam mass flow out from the superheater*

V_s - *superheater volume*

ρ_s - *superheated steam density*

w_a - *attemporator water mass flow*

Friction loss *(momentum balance)*

$$p_v - p_s = \frac{w_v^2}{\rho_v} f_s$$

where :

p_v - *steam drum pressure*

p_s - *superheated steam pressure*

ρ_v - *saturated steam density (from the drum model)*

f_s - *friction coefficient*

Gas-tube heat balance

$$Q_{gs} = Q_s + M_s C_{st} \frac{d}{dt}\left(T_{st}\right)$$

where :

Q_{gs} - *heat supplied to the superheater (from the furnace model)*

Q_s - *heat transferred to the steam*

M_s - *mass of superheater tubes*

C_{st} - *heat capacitance of superheater tubes*

T_{st} - *metal tube temperature*

Heat transfer tube-steam

$$Q_s = k_s w_v^{0.8}\left(T_{st} - T_s\right)$$

where :

k_s - *an experimental coefficient*

T_s - *steam temperature*

Heat balance for steam

$$Q_s + w_v h_v = w_s h_s - \left(h_a - h_f\right) w_a + V_s \frac{d}{dt}\left(\rho_s h_s\right)$$

where :

h_v - *specific enthalpy of saturated steam (from the drum)*

h_s - *specific enthalpy of superheated steam*

h_a - *specific enthalpy of attemporation water*

h_f - *specific enthalpy of evaporation*

Superheat steam temperature

$$T_s = \frac{h_s - h_{ref}}{c_{ps}} + T_{ref}$$

where :

h_{ref} - *reference steam enthalpy condition*

T_{ref} - *reference steam temperature condition*

c_{ps} - *specific heat of steam at constant pressure*

Superheat steam pressure

$$p_s = R_s \, \rho_s \, T_s$$

where:

p_s - *superheat steam pressure*

R_s - *ideal gas constant*

Input variables to the superheater and attemporator module :

w - *attemporator water mass flow.*

h_a - *attemporator water specific enthalpy*

Incoming variables from other modules of the boiler:

p_v - *steam drum pressure (from the drum model)*

ρ_v - *saturated steam density (from the drum model)*

Q_{gs} - *heat supplied to the superheater (from the furnace model)*

h_v - *specific enthalpy of saturated steam (from the drum model)*

The steam mass flow out from the superheater (w_s) represents an interconnection between the boiler model and the steam turbine model. This flow, w_s, is calculated in the steam turbine model.

Outputs from the superheater model connected to other modules of the boiler:

w_v - *steam mass flow from the drum (to the drum model)*

T_{st} - *metal tube temperature (to the furnace model)*

Outputs from the superheater model which are also outputs from the boiler model :

p_s - *superheated steam pressure*

T_s - *superheated steam temperature*

The Reheater

The reheater equations are very similar to the superheater equations. The only difference is that an attemporation flow is not considered. The equations are:

Mass balance

$$w_{ri} - w_{ro} = V_r \frac{d}{dt}\left(\rho_{rh}\right)$$

where :

w_{ri} - *reheater inlet steam mass flow*

w_{ro} - *reheater outlet steam mass flow*

V_r - *reheater volume*

ρ_{rh} - *reheated steam density*

Reheat steam temperature

$$T_r = \frac{h_{ro} - h_{ref}}{c_{pr}} + T_{ref}$$

where :

h_{ref} - *reference steam enthalpy condition*

T_{ref} - *reference steam temperature condition*

c_{pr} - *specific heat of steam at constant pressure*

Reheat steam pressure

$$p_{ro} = R_r \, \rho_{rh} \, T_r$$

where:

p_{ro} - *reheat steam pressure*

R_r - *ideal gas constant*

Reheater outlet mass flow

$$\dot{w}_{ro} = \left(w_{ri} - w_{ro}\right) / \tau_{rh}$$

where : τ_{rh} - an empirical flow time constant

Gas-tube heat balance

$$Q_{rs} = Q_{rh} + M_r C_{rh} \frac{d}{dt}\left(T_{rh}\right)$$

where :

Q_{rs} - *heat supplied to the reheater*

Q_{rh} - *heat transferred to the steam*

M_r - *mass of reheater tubes*

C_{rh} - *heat capacitance of the tubes*

T_{rh} - *metal tube temperature*

Heat transfer tube-steam

$$Q_{rh} = K_{rh} w_{ri}^{0.8}\left(T_{rh} - T_r\right)$$

where :

K_{rh} - *an experimental coefficient*

T_r - *reheated steam temperature*

Heat balance for steam

$$Q_{rh} + w_{ri} h_{ri} = w_{ro} h_{ro} + V_r \frac{d}{dt}\left(\rho_{rh} h_{ro}\right)$$

where :

h_{ri} - *inlet steam specific enthalpy*

h_{ro} - *outlet steam specific enthalpy*

Inputs to the reheater model :

p_{ri} - *inlet steam pressure (from the steam turbine)*

w_{ro} - *outlet steam mass flow (from the steam turbine)*

T_{ri} - *inlet steam temperature (from the steam turbine)*

Input internal to the boiler model:

Q_{rs} - *heat supplied to the reheater (from the furnace model)*

Output from the reheater which is internal to the boiler model:

T_{rh} - *reheater metal tube temperature*

Outputs from the reheater which are also outputs from the boiler model:

p_{ro} - *outlet steam pressure*

T_r - *outlet steam temperature*

w_{ro} - *outlet steam mass flow (to the steam turbine)*

4.3.1.5 Block Diagram

The block diagram of the boiler model is presented in Fig. 4.1 This is an overall block diagram that shows interconnections between modules in the boiler model. Each individual module is given according to the equations listed in the next section.

4.3.1.6 Model structure for implementation as an algorithm

The equations presented in Section 4.3.1.4 are now integrated into an implementable algorithm of the boiler module. To do this, certain modifications are required. In particular, all the algebraic equations are rearranged and re-ordered in such a way that all variables on the right hand side are known at the time of calculation. The differential equations follow the algebraic equations.

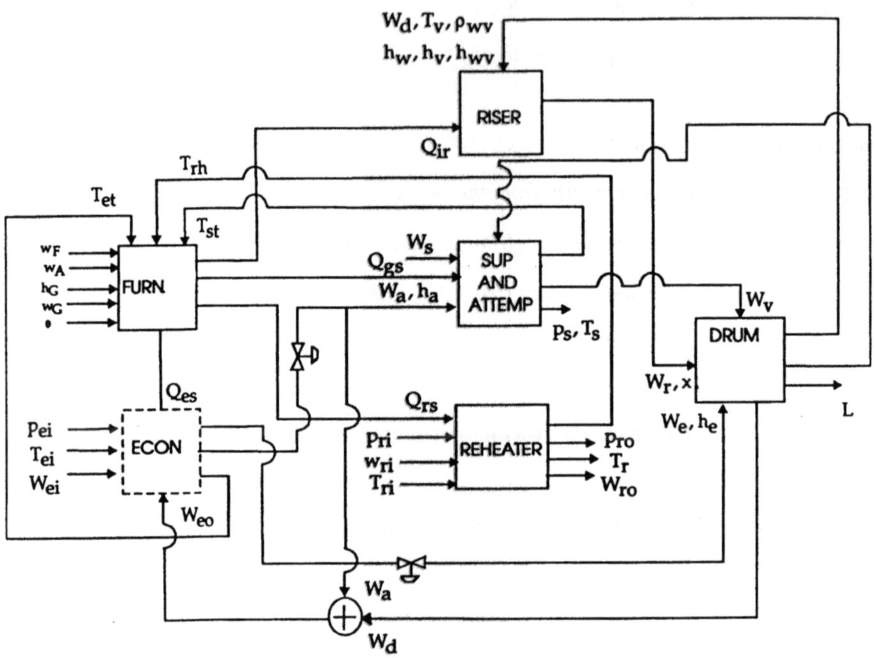

Fig. 4.1 Boiler - block diagram.

The algorithm may be written using a variety of high level languages, for example, FORTRAN, C, MATLAB or $MATRIX_X$.

<u>Furnace</u>

- *Inputs* : w_F, w_A, h_G, w_G, θ, T_{st}, T_{rh}, T_{et}, h_A
- *Parameters* : k_F, k, k_{gs}, c_{gs}, k_{rs}, V_F, C_F, R_s γ, k_{es}, h_{ref}, T_{ref}
- *States* : ρ_{EG}, x_{F1}
- *Algebraic equations*:

$$h_{EG} = \frac{x_{F1}}{\rho_{EG}}$$

$$T_g = \frac{h_{EG} - h_{ref}}{c_{pg}} + T_{ref}$$

$$p_G = R_{EG}\,\rho_{EG}\,T_g$$

$$w_{EG} = k_F p_G$$

$$Q_{ir} = \theta k V_F \sigma T_g^4$$

$$Q_{is} = (1-\theta) k V_F \sigma T_g^4$$

$$Q_{gs} = Q_{is} + k_{gs} w_{EG}^{0.6}(T_g - T_{st})$$

$$T_{gr} = T_g + \frac{1}{c_{gs}} \frac{1}{w_{EG}} (Q_{is} - Q_{gs})$$

$$Q_{rs} = k_{rs} w_{EG}^{0.6}(T_{gr} - T_{rh})$$

$$T_{ge} = T_{gr} - \frac{1}{c_{gs}} \frac{1}{w_{EG}} Q_{rs}$$

$$Q_{es} = k_{es} w_{EG}^{0.6}(T_{ge} - T_{et})$$

$$T_{g1} = T_{ge} - \frac{1}{c_{gs}} \frac{1}{w_{EG}} Q_{es}$$

$$y = 100(w_A + \gamma w_G - w_F R_s)\frac{1}{w_F R_s}$$

◊ *Differential equations:*

$$\frac{d}{dt} x_{FI} = \frac{1}{V_F}\left(C_F w_F + h_A w_A + h_G w_G - Q_{ir} - Q_{is} - w_{EG} R_s \left(1 + \frac{y}{100}\right) h_{EG}\right)$$

$$\frac{d}{dt} \rho_{EG} = \frac{1}{V_F}\left(w_F + w_A + w_G - w_{EG}\right)$$

◊ *Outputs*: Q_{ir}, Q_{is}, Q_{rs}, Q_{es}, p_G

Risers

◊ *Inputs* : $w_d, Q_{ir}, h_w, h_v, h_{wv}, T_v, \rho_{wv}, \rho_v$

◊ *Parameters* : $K_r, V_r, M_r, C_{rt}, \tau_r$

◊ *States* : ρ_r, h_r, T_{rt}, w_r

◊ *Algebraic equations*:

$$x = \frac{h_r - h_{wv}}{h_v - h_{wv}}$$

$$\rho_r = \left[\frac{x}{\rho_v} + \frac{(1-x)}{\rho_{wv}}\right]^{-1}$$

$$Q_r = k_r (T_{rt} - T_v)^3$$

◊ *Differential equations*:

$$\frac{d}{dt} h_r = \frac{1}{\rho_r V_r}\left(w_d h_w - w_r h_r + Q_r\right)$$

$$\frac{d}{dt} T_{rt} = \frac{1}{M_r C_{rt}}\left(Q_{ir} - Q_r\right)$$

$$\frac{d}{dt} w_r = \frac{1}{\tau_r}\left(w_d - w_r\right)$$

◊ *Outputs*: w_r, x

Drum:

◊ *Inputs* : $w_e, x, w_r, w_v, h_e, v_{dow}$

◊ *Parameters* : V, k_{ec}, r

◊ *States* : m_{dL}, x_{D1}, x_{D2}

◊ *Algebraic equations*:

$$h_w = \frac{x_{D1}}{m_{dL}}$$

$$p_w = f_3(h_w) \qquad \text{[saturated water tables]}$$

$\rho_w = f_4(h_w)$ [saturated water tables]

$T_w = f_5(h_w)$ [saturated water tables]

$w_d = v_{dow}\ \rho_w$

$$V_L = \frac{m_{dL}}{\rho_w}$$

$V_v = V - V_L$

$$\rho_v = \frac{x_{D2}}{V_v}$$

$h_v = f_5(\rho_v)$ [saturated steam tables]

$T_v = f_7(\rho_v)$ [saturated steam tables]

$p_v = f_8(\rho_v)$ [saturated steam tables]

$h_{wv} = f_9(\rho_v)$ [saturated steam tables]

$W_{ec} = k_{ec}(T_w - T_v)$

$L = f_{10}^{-1}(V_L)$

$$f_{10}(L) = \tfrac{1}{3}\pi L^2(3r - L) + \tfrac{1}{2}(W - 2r)r^2(\theta - \sin\theta)$$

$$\textit{where} \quad \theta = 2\cos^{-1}\left(\frac{r-L}{r}\right)$$

◊ *Differential Equations*:

$$\frac{d}{dt} m_{d1} = w_e + (1-x)w_r - w_d - w_{ec}$$

$$\frac{d}{dt} x_{D1} = w_e h_e + (1-x)w_r h_{wv} - w_d h_w - w_{ec} h_v$$

$$\frac{d}{dt} x_{D2} = w_{ec} + x w_r - w_v$$

◊ *Outputs*: p_v, p_w, ρ_{wv}, ρ_v, h_v, h_w, h_{wv}, w_d

Superheater and Attemporator:

- *Inputs* : $w_a, w_s, p_v, \rho_v, Q_{gs}, h_v, h_a$
- *Parameters* : $f_s, k_s, V_s, M_s, C_{st}, c_{ps}, T_{ref}, h_{ref}$
- *States* : p_s, T_{st}, x_{s1}
- *Algebraic equations*:

$$h_s = \frac{x_{s1}}{\rho_s}$$

$$T_s = \frac{h_s - h_{ref}}{c_{ps}} + T_{ref}$$

$$p_s = R_s \rho_s T_s$$

$$h_f = f_{13}(p_s) \qquad \text{[saturated steam tables]}$$

$$w_v = \sqrt{\frac{(p_v - p_s)\rho_v}{f_s}}$$

$$Q_s = K_s w_v^{0.8}(T_{st} - T_s)$$

- *Differential Equations*:

$$\frac{d}{dt}\rho_s = \frac{1}{V_s}(W_v - W_s)$$

$$\frac{d}{dt}T_{st} = \frac{1}{M_s C_{st}}(Q_{gs} - Q_s)$$

$$\frac{d}{dt}x_{s1} = \frac{1}{V_s}\left[Q_s + w_v h_v - w_s h_s + (h_a - h_f)w_a\right]$$

- *Outputs*: W_v, T_{st}, P_s, T_s

Reheater

- *Inputs* : P_{ri}, w_{ri}, T_{ri}, Q_{rs}, ρ_{ri}, h_{ri}
- *Parameters :* τ_r, k_{rh}, V_{rh}, M_r, C_{rh}, c_{pr}, T_{ref}, h_{ref}
- *States* : ρ_{rh}, T_{rh}, x_{RH1}, w_{ro}
- *Algebraic equations*:

$$h_{ro} = \frac{x_{RH1}}{\rho_{rh}}$$

$$T_r = \frac{h_{ro} - h_{ref}}{c_{pr}} + T_{ref}$$

$$p_{ro} = R_r \, \rho_{rh} \, T_r$$

$$Q_{rh} = K_{rh} \, w_{ri}^{0.8} (T_{rh} - T_r)$$

- *Differential Equations*:

$$\frac{d}{dt} \rho_{rh} = \frac{1}{V_{rn}} (W_{ri} - W_{ro})$$

$$\frac{d}{dt} T_{rh} = \frac{1}{M_r C_{rh}} (Q_{rs} - Q_{rh})$$

$$\frac{d}{dt} x_{RH1} = \frac{1}{V_{rh}} \left(Q_{rh} + w_{ri} h_{ri} - w_{ro} h_{ro} \right)$$

$$\frac{d}{dt} w_{ro} = \left(w_{ri} - w_{ro} \right) / \tau_{rh}$$

- *Outputs*: T_{rh}, P_{ro}, T_r , w_{ro}

4.3.2 Gas Turbine

4.3.2.1 Introduction

The main components of an industrial power plant gas turbine are the compressor, the combustion chamber and the turbine. The gas turbine operates as follows.

Air at atmospheric pressure enters the gas turbine at the compressor inlet. After compression of the air to achieve the most favourable conditions for combustion, fuel gas is mixed with the air in the combustion chamber, combustion takes place and the hot exhaust gases are expanded through the turbine to produce mechanical power. In terms of energy conversion, chemical energy present in the combustion reactants is transferred to the gas stream during combustion. This energy - measured in terms of gas enthalpy - is then converted into mechanical work, by expanding the gas through the turbine. Thus the excess mechanical power available for application elsewhere, after accounting for the power required to drive the compressor, is derived ultimately from the combustion process. Without combustion, assuming 100% efficient compressor and turbine operation, the power developed by the turbine would be exactly matched by the power required to drive the compressor.

4.3.2.2 Literature Review

A thorough introduction to gas turbine theory is provided in the book of [*Cohen, et al., 1987*]. There also exists a large literature on the modelling of gas turbines. Model complexity varies according to the intended application. Detailed first principles modelling, based upon fundamental mass, momentum and energy balances is reported by [*Fawke, et al., 1972*], [*Shobeiri, 1987*], [*Onions and Foss, 1986*]. These models describe the spatially distributed nature of the gas flow dynamics by dividing the gas turbine into a number of sections. Throughout each section, the thermodynamic state is assumed to be constant with respect to location, but varying with respect to time. Mathematically, the full partial differential equation model description is reduced to a set of ordinary differential equations which facilitate easier application within a computer simulation program. For a detailed model, a section might consist of a single compressor or turbine stage. Much simpler models result if the gas turbine is decomposed into just three sections corresponding to the main turbine components i.e. compressor, combustor and turbine [*Hussain and Seifi, 1992*].

Instead of applying the fundamental conservation equations, as described above, another modelling approach is to characterise gas turbine performance by utilizing real steady state engine performance data [*Hung, 1991*], [*Biss and Grimble, 1993*]. It is assumed that transient thermodynamic and flow processes are characterised by a continuous progression along the steady state performance curves, this is known as the *quasi-static* assumption. The dynamics of the gas turbine, e.g. combustion delay, rotor inertia, fuel pump lag etc. are then represented as lumped quantities separate from the steady-state performance curves. Very simple models result if it is further assumed that the gas turbine is operated at all times close to rated speed [*Rowen, 1983*]. In this case, the modelling of gas turbine performance over a wide range of operating conditions, as is experienced during start up or shutdown procedures, is not attempted.

For the purposes of developing a gas turbine model suitable for the evaluation of supervisory CC/CHP control strategies, it is necessary to capture the main dynamics of the gas turbine over a wide range of operating conditions. Consequently, the model described in this report retains the structure of [*Rowen, 1983*] but extends the model's range of validity. This is achieved by including a significantly more complicated representation of the compressor, combustion chamber and turbine components. As in [*Hussain and Seifi, 1992*], the compressor and turbine components are replaced by nozzle equivalents. The one-dimensional equations for a uniform polytropic compression or expansion are then applied. Furthermore the production of the gas pollutants NO_X and CO by the combustion process is modelled using data from [*Thermoflow, Inc., 1991*].

4.3.2.3 Model Assumptions

The main modelling assumptions are as follows:

(i) air and combustion products are treated as perfect gases

(ii) specific heats are assumed constant for combustion products, air and injected steam

(iii) flow through nozzles is described by a one dimensional adiabatic uniform polytropic process

(iv) energy storage and transport delay in the compressor, turbine andcombustion chamber are relatively small, thus steady state equations are applied

(v) inlet kinetic energy of gas flows into the compressor and turbine are treated as negligible

(vi) air mass flow through the compressor is controllable via inlet guide vanes

(vii) N_{ox}, CO pollutant formation is modelled as a static function of steam/fuel ratio.

4.3.2.4 Gas Turbine Equations

The gas turbine is divided into four interconnected subsystems. They are:

- fuel system (fuel valve with actuator)
- compressor
- combustor
- turbine

Fuel System

The fuel system consist of the fuel valve and the actuator. The fuel flow out from the fuel systems results from the inertia of the fuel system actuator and of the valve positioner:

Fuel system actuator (In Laplace - transform form)

$$w_f = \frac{k_{ff}}{\tau_f s + 1} e_1$$

where :

w_f -	*fuel mass flow*
k_{ff} -	*fuel system gain constant*
τ_f -	*fuel system time constant,*
e_1 -	*valve position.*

Valve positioner (in Laplace - transform form):

$$e_1 = \frac{a}{bs + c} e_2$$

where :

a,b,c - *valve parameters*

e_2 - *internal signal*

Signal for valve positioner (In Laplace-transform form)

$$e_2 = MF - k_f w_f + F_d \omega e^{-s\tau}$$

where :

MF - *minimum fuel signal*

k_f *feedback coefficient*

F_d - *fuel demand signal*

ω - *rotation speed of the turbine*

τ - *fuel system pure time delay*

Input variables to the fuel system :

F_d - *fuel demand signal (from a controller)*

ω - *rotation speed of the turbine (from the generator model)*

Output variable from the fuel system model :

w_f - *fuel flow to the combustor*

Compressor

The compressor is described by the following equations:

One dimensional steady flow nozzle equation for a uniform polytropic compression *[Knowles, 1990 ; Rogers and Mayhew 1992 ; Cohen et al., 1987].*

$$w_a = A_o \left[\left(\frac{2m_a}{\eta_{\infty c}(m_a - 1)} \right) \rho_i p_{cin} \left(r_c^{2/m_a} - r_c^{(m_a+1)/m_a} \right) \right]^{1/2}$$

where:

w_a- *air mass flow into the compressor*

A_o - *compressor exit flow area*

$\eta_{\infty c}$ - *compressor polytropic efficiency*

ρ_i - *inlet air density*

p_{cin} - *inlet air pressure*

m_a - *polytropic index*

r_c - *pressure ratio (outlet/inlet)*

Polytropic index equation

$$m_a = \frac{\gamma_a}{\gamma_a - \frac{(\gamma_a - 1)}{\eta_{\infty c}}}$$

where :

γ_a - $= (c_{pa}/c_{va})$ *ratio of specific heats for air (constant)*

c_{pa} - *specific heat at constant pressure for air*

c_{va} - *specific heat at constant volume for air*

Outlet air pressure equation

$$p_{cout} = p_{cin}\, r_c$$

Outlet air temperature equation

$$\left(T_{cout} / T_{cin}\right) = r_c^{\left(\frac{\gamma_{a-1}}{\gamma_a \eta_{\infty c}}\right)}$$

where:

T_{cout} - *outlet air temperature*

T_{cin} - *inlet air temperature*

Compressor power consumption equation

$$P_c = \frac{w_{ain} \Delta h_I}{\eta_c \eta_{trans}}$$

where:

P_c - *compressor power consumption*

Δh_I - *isentropic enthalpy change corresponding to a compression from p_{cin} to p_{cout}*

η_c - *overall compressor efficiency*

η_{trans} - *transmission efficiency from turbine to compressor*

Overall compressor efficiency equation

$$\eta_c = \frac{1-(r_c)^{\left(\frac{\gamma_a-1}{\gamma_a}\right)}}{1-(r_c)^{\left(\frac{\gamma_a-1}{\gamma_a\eta_{\infty c}}\right)}}$$

Perfect gas isentropic enthalpy change equation

$$\Delta h_I = c_{pa} T_{cin} (r_c^{R_a/c_{pa}} - 1)$$

where :

c_{pa} - *specific heat of air at constant pressure*

R_a - *air gas constant*

Combustion Chamber

Energy and pressure drop equations adapted from [*Cohen, Rogers and Saravanamutto 1987*] to include steam injection effects.

Exhaust gas mass flow

$$w_G = w_a = w_f + w_{is}$$

Combustion energy equation

$$w_G c_{pg}(T_{Tin} - 298) + w_f \Delta h_{25}$$
$$+ w_a c p_a(298 - T_{cout}) + w_{is} c_{ps}(298 - T_{is}) = 0$$

where :

w_f - *fuel mass flow*
w_{is} - *injection steam mass flow*
c_{pg} - *specific heat of combustion gases (constant*
T_{Tin} - *Turbine inlet gas temperature*
Δh_{25} - *Specific enthalpy of reaction at reference temperature of 25°c*
c_{ps} - *specific heat of steam (constant)*
T_{is} - *temperature of insected steam*

Combustion chamber pressure loss

$$p_{Tin} = p_{cout} - \Delta p$$

with

$$\Delta p = p_{cout}\left[\left(k_1 + k_2\left(\frac{T_{Tin}}{T_{cout}} - 1\right)\right)\frac{R}{2}\left(\frac{w_G}{A_m p_{cout}}\right)^2 T_{cout}\right]$$

where:

p_{Tin} - *pressure of combustion gases at turbine inlet*

Δp - *combustion chamber pressure loss*

k_1, k_2 - *pressure loss coefficients*

R_{cg} - *universal gas constant for combustion gases*

A_m - *combustion chamber mean cross-sectional area*

Pollutant formation

$$g'_{cnox} = f_{GT2}\left(\frac{w_{is}}{w_f}\right)$$

$$g'_{cco} = f_{GT3}\left(\frac{w_{is}}{w_f}\right)$$

where :

g'_{cnox} - *mass flow of NO_x*

g'_{cco} - *mass flow of CO*

f_{GT2} - *experimental curve : NO_x mass flow as a function of steam to fuel mass flow ratio*

f_{GT3} - *experimental curve : CO mass flow as a function of steam to fuelmass flow ratio*

Pollutant formation measurement dynamics

$$g_{cnox}(t) = g'_{cnox}(t - \tau_m)$$

$$g_{cco}(t) = g'_{cco}(t - \tau_m)$$

where :

$g_{cnox}(t)$ - *delayed (measured) NO_x mass flow*

$g_{cco}(t)$ - *delayed (measured) CO mass flow*

τ_m- *measurement delay*

<u>The turbine</u>

Temperature - pressure relationship

$$\left(\frac{T_{Tout}}{T_{Tin}}\right) = r_T^{\eta_{\infty T}\left(\frac{\gamma_{cg}-1}{\gamma_{cg}}\right)}$$

where :

T_{Tout} - *gas temperature at exit of turbine*

r_T - $=\left(\frac{P_{Tout}}{P_{Tin}}\right)$ *outlet to inlet turbine pressure ratio*

$\eta_{\infty T}$ - *turbine polytropic efficiency*

γ_{cg} - $=\left(\frac{c_{pg}}{c_{vg}}\right)$ *ratio of specific heats for combustion gases*

Gas mass flow through the turbine:

$$w_G = A_{To}\left[\left(\frac{2\eta_{\infty T} m_{cg}}{m_{cg}-1}\right)\rho_{Tin} p_{Tin}\left(r_T^{(2/m_{cg})} - r_T^{\left(\frac{m_{cg}+1}{m_{cg}}\right)}\right)\right]^{1/2}$$

with

$$m_{cg} = \frac{\gamma_{cg}}{\gamma_{cg} - \eta_{\infty T}(\gamma_{cg}-1)}$$

$$\rho_{Tin} = f_{GT4}(T_{Tin}, P_{Tin})$$

where :

m_{cg} - *combustion gases polytropic index*

ρ_{Tin} - *inlet gas density*

f_{GT4} - *gas tables function*

w_G - *turbine gas mass flow*

Overall turbine efficiency

$$\eta_T = \frac{1-(r_T)^{\left(\frac{\eta_{\infty T}(\gamma_{cg}-1)}{\gamma_{cg}}\right)}}{1-(r_T)^{\left(\frac{\gamma_{cg}-1}{\gamma_{cg}}\right)}}$$

where : η_T - *overall turbine efficiency*

Power delivery

$$P_T = \eta_T w_g \Delta h_I$$

$$P_{mech} = P_T - P_c$$

with

$$\Delta h_I = c_{pg} T_{Tin}\left(r_T^{R_{cg}/c_{pg}} - 1\right)$$

where :

P_T - *mechanical power delivered by turbine*

P_c - *power required to drive the compressor*

P_{mech} - *net available mechanical power*

Δh_I - *isentropic enthalpy change for a gas expansion from P_{Tin} to P_{Tout}.*

4.3.2.5 Block Diagram

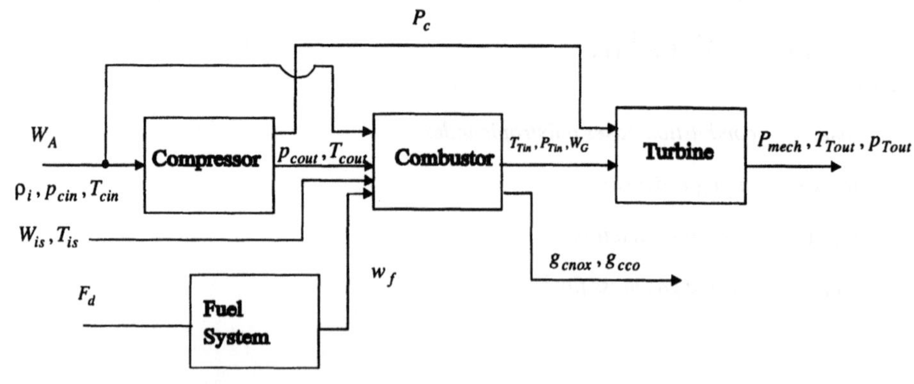

Fig. 4.2 Gas turbine block diagram.

4.3.2.6 Model structure for implementation as an algorithm

Fuel system

The following block diagram is implemented.

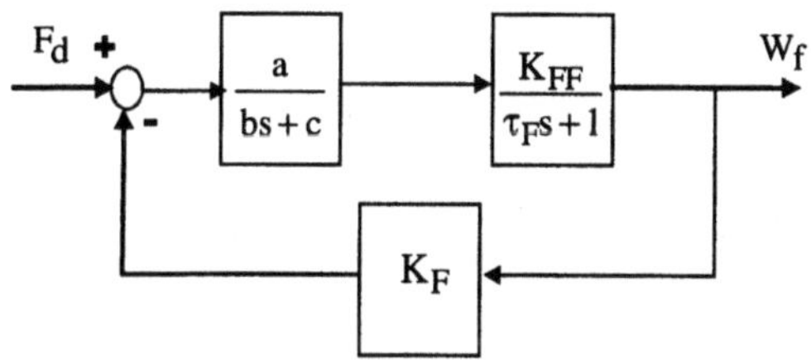

Fig. 4.3 Fuel system - block diagram.

- *Inputs* : F_d,
- *Parameters* : a, b, c, τ_F, K_F, K_{FF}
- *Output* : w_f

Compressor

- *Inputs* : w_a, ρ_i, p_{cin}, T_{cin}
- *Parameters :* $\eta_{\infty c}$, A_o, η_{trans}, τ_{cD}, c_{pair},
- *Constants :* γ_a, R_{air}
- *Algebraic equations:*

$$m_a = \frac{\gamma_a}{\gamma_a - (\gamma_a - 1)/\eta_{\infty c}}$$

$$r_c^{\left(\frac{2}{m_a}\right)} - r_c^{\left(\frac{m_a+1}{m_a}\right)} = \frac{\eta_{\infty c}(m_a - 1)}{2m_a} \frac{1}{\rho_i p_{cin}} \left(\frac{w_a}{A_o}\right)^2$$

Solve for r_c

$p_{cout} = p_{cin} r_c$

$$T_{cout} = T_{cin} r_c^{\left(\frac{\gamma_a - 1}{\gamma_a \eta_{\infty c}}\right)}$$

$$\eta_c = \frac{1-(r_c)^{\left(\frac{\gamma_a-1}{\gamma_a}\right)}}{1-(r_c)^{\left(\frac{\gamma_a-1}{\gamma_a \eta_{\infty c}}\right)}}$$

$$\Delta h_I = c_{pair} T_{cin} \left(r_c^{\frac{R_{air}}{c_{pair}}} - 1\right)$$

$$P_c = \frac{w_a \Delta h_I}{\eta_c \eta_{trans}}$$

◊ *Outputs*: p_{cout}, T_{cout}, P_c

Combustion chamber

◊ *Inputs* : p_{cout}, T_{cout}, w_a, w_f, w_{is}, T_{is}

◊ *Parameters* : c_{pg}, c_{pa}, c_{ps}, Δh_{25}, k_1, k_2, R_{cg}, A_m, τ_m, h_{ref}, T_{ref}

◊ *Constants* : R_{cg}

◊ *Algebraic equations:*

$$w_G = w_a + w_f + w_{is}$$

$$T_{Tin} = \frac{1}{w_G} \frac{1}{c_{pg}} \left[w_a c_{pa} (T_{cout} - 298) + w_{is} c_{ps} (T_{is} - 298) - w_f \Delta h_{25} \right]$$

$$\Delta p = p_{cout} \left[\left(k_1 + k_2 \left(\frac{T_{Tin}}{T_{cout}} - 1 \right) \right) \frac{R_{cg}}{2} \left(\frac{w_G}{A_m p_{cout}} \right)^2 T_{cout} \right]$$

$$p_{Tin} = p_{cout} - \Delta p$$

$$g'_{cnox} = f_{GT2}(\frac{w_{is}}{w_f})$$

[experimental curve : NO_x content as a function of steam to fuel ratio]

$$g'_{cco} = f_{GT3}(\frac{w_{is}}{w_f})$$

[experimental curve : CO content as a function of steam to fuel ratio]

$$h_{Tin} = h_{ref} + c_{pg}(T_{Tin} - T_{ref})$$

◊ *Dynamics*:

$$g_{cnox}(t) = g'_{cnox}(t - \tau_m)$$

$$g_{cco}(t) = g'_{cco}(t - \tau_m)$$

◊ *Outputs:* W_G, T_{Tin}, p_{Tin}, g_{cnox}, g_{cco}

The turbine

◊ *Inputs* : T_{Tin}, p_{Tin}, P_c, W_G

◊ *Parameters :* p_{Tout}, $\eta_{\infty t}$, A_{to}, c_{pg},

◊ *Constants :* γ_{cg}, R_{cg}

◊ *Algebraic equations:*

$$m_{cg} = \frac{\gamma_{cg}}{\gamma_{cg} - \eta_{\infty T}(\gamma_{cg} - 1)}$$

$$\rho_{Tin} = \frac{p_{Tin}}{R_{cg} T_{Tin}}]$$

$$r_T^{(\frac{2}{m_{cg}})} - r_T^{(\frac{m_{cg}+1}{m_{cg}})} = \left(\frac{W_G}{A_{To}}\right)^2 \frac{(m_{cg} - 1)}{2\eta_{\infty T} m_{cg} \rho_{Tin} p_{Tin}}$$

Solve for r_T

$$p_{Tout} = r_T p_{Tin}$$

$$T_{Tout} = T_{Tin} r_T^{\frac{\eta_{\infty T}(\gamma_{cg} - 1)}{\gamma_{cg}}}$$

$$\Delta h_I = c_{pg} T_{Tin} \left(r_T^{R_{cg}/c_{pg}} - 1 \right)$$

$$\eta_T = \frac{1-(r_T)^{(\eta_{\infty T}\frac{\gamma_{cg}-1}{\gamma_{cg}})}}{1-(r_T)^{(\frac{\gamma_{cg}-1}{\gamma_{cg}})}}$$

$$p_T = \eta_T W_G \Delta h_I$$

$$h_{Tout} = h_{Tin} + \eta_T \Delta h_I$$

$$P_{mech} = p_T - p_c$$

◊ *Outputs:* P_{mech} T_{Tout}, P_{Tout}, h_{Tout}

4.3.3 Steam Turbine

4.3.3.1 Introduction

As in the case of the gas turbine, mechanical power is obtained by removing energy from the gas stream by expanding it to a lower pressure. In the case of the tandem compound single reheat turbine, steam enters the High Pressure chest via the main admission control valves. After passing through the HP stage it is circulated to the boiler to be reheated at constant pressure. Then the steam passes through the Intermediate Pressure (IP) stage, crossover storage and finally through the Low Pressure (LP) stage before returning to the condenser. This configuration has been selected to be modelled because it includes all the essential components for typical steam turbine configurations whilst remaining relatively simple.

4.3.3.2 Literature Review

A quite detailed steam turbine model based upon fundamental balances is presented in [*Ray, 1980*]. Ray developed equations that are applicable to the individual impulse and reaction stages of the turbine. This model has been applied in designing multivariable controllers for steam turbines. Much simpler, essentially linear models have also been reported for application to electrical network transient stability studies [*IEEE*, 1973 ; 1991]. For the development of supervisory control strategies an intermediate solution in terms of complexity is sought where all the primary inputs and outputs and the salient dynamics are represented. Such a model has been constructed by retaining the basic structure of the IEEE model whilst including some of the more important non-linearities associated with the turbine. Similar to the gas turbine model, the turbine is replaced by equivalent nozzles representing the high pressure, intermediate pressure and low pressure stages. Energy and mass balances are applied to these sections.

2.3.3 ***Model Assumptions***

The main modelling assumptions are as follows:

(i) Superheated steam is treated as an ideal gas.

(ii) HP,IP,LP, turbine stages are converted to equivalent nozzles through which one-dimensional uniform polytropic steam expansion takes place.

(iii) Energy storage volumes are lumped.

(iv) Inlet kinetic energy of steam to each stage is negligible.

(v) Mass flow dynamics between input and output modelled as a 1st order lag.

4.3.3.4 Steam Turbine Equations

The steam turbine model is sub-divided into three sections. A section is defined here as consisting of a lumped steam storage volume having dynamics, followed by a complete turbine stage which is modelled by steady-state relationships. [*Knowles, 1990*] justifies the use of steady-state relationships for the turbine stages, because energy storage and transport delay effects are relatively small compared to other power system components.

Note that a complete turbine stage (high pressure, intermediate pressure or low pressure) comprises of a number of impulse and/or reaction stages in series. The three sections relating to the tandem-compound single-reheat steam turbine is the actual configuration modelled. However any other desired steam turbine configuration can easily be constructed from the steam storage and turbine stage building blocks.

High pressure section

Mass balance for steam

$$w_i - w_{ohp} = V\frac{d}{dt}(\rho_o)$$

where :

w_i - *inlet steam flow*

w_{ohp} - *outlet steam flow*

V - *chest storage volume*

ρ_o - *outlet steam density from the chest*

Section mass flow dynamics

$$\dot{w}_{ouhp} = \frac{1}{\tau_{hp}}\left(w_{in} - w_{ouhp}\right)$$

where:

τ_{hp} - *an empirical HP section time constant*

Heat balance

$$w_i h_i - w_{ohp} h_o = V\frac{d}{dt}(\rho_o h_o)$$

where :

h_i - *inlet steam specific enthalpy*

h_o - *outlet steam specific enthalpy (from the steam chest)*

Ideal gas relations

$$T_o = \frac{h_o - h_{in}}{c_p} + T_{in}$$

$$p_o = R\,\rho_o\,T_o$$

where :

T_o - *outlet steam temperature (from the steam chest)*

c_p - *specific heat at constant pressure*

Nozzle equation

$$r_{hp}^{(\frac{2}{m})} - r_{hp}^{(\frac{m+1}{m})} = \frac{w_{ohp}^2}{A_{hp}\rho_o p_o}\left(\frac{m-1}{2\eta_{\infty hp} m}\right)$$

where :

$$r_{hp} = \frac{p_{ohp}}{p_o}$$

$$m = \frac{\gamma}{\gamma - \eta_{\infty hp}(\gamma - 1)}$$

$$\gamma = \frac{c_p}{c_v}$$

c_p - *specific heat at constant pressure*

c_v - *specific heat at constant volume*

p_{ohp} - *outlet pressure from the reaction*

$\eta_{\infty hp}$- *polytropic efficiency of the section*

A_{hp} - *cross-sectional exit area*

Uniform polytropic expansion - temperature ratio

$$\frac{T_{ohp}}{T_o} = \left(\frac{p_{ohp}}{p_o}\right)^{\eta_{\infty hp}(\frac{\gamma-1}{\gamma})}$$

where : T_{ohp} - *steam temperature at the outlet from the section*

Enthalpy change for isentropic expansion of perfect gas

$$\Delta h_{Ihp} = c_p T_o \left(r_{hp}^{R/c_p} - 1 \right)$$

where :

Δh_{Ihp} - *isentropic enthalpy change when pressure changes from p_o to p_{ohp}*

h_{ohp} - *specific enthalpy of the outlet steam (from the section)*

Overall section efficiency

$$\eta_{hp} = \frac{1 - r^{(\eta_{\infty hp} \frac{\gamma - 1}{\gamma})}}{1 - r^{(\frac{\gamma - 1}{\gamma})}}$$

where : η_{hp} - *overall efficiency of the section*

Power developed by the section

$$P_{hp} = \eta_{hp} \Delta h_{Ihp}$$

Outlet enthalpy

$$h_{ou} = h_o + c_p (T_{ou} - T_o)$$

Intermediate pressure section

The model is the same as for HP section, except that there is no steam storage since this exists in the boiler model as the reheater.

Low pressure section

The model is the same as for HP section.

Total power delivered

$$P_{st} = P_{hp} + P_{ip} + P_{lp}$$

4.3.3.5 Block diagram

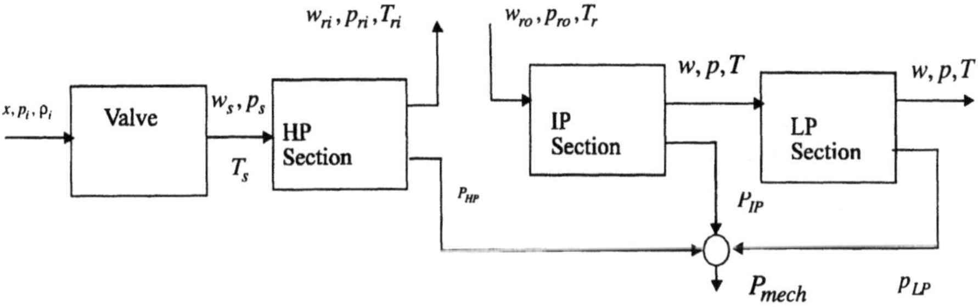

Fig. 4.4 Steam turbine - block diagram.

4.3.3.6 Model Structure for Implementation as an Algorithm

The steam turbine model consists of three sections:

- High pressure section
- Intermediate pressure section
- Low pressure section

For the HP and LP sections the same set of equations apply. The differences occur only in the input signals and parameters. For the IP section there are no storage dynamics represented, because this is modelled in the reheater, which is a part of the boiler model.

In the following a generic turbine section algorithm is presented, with definitions of inputs, outputs and parameters given for the LP, IP and HP sections. The equations that do not apply to the IP section will be highlighted.

<u>Turbine section</u>

◊ *Inputs* : $w_{in}, p_{in}, h_{in}, T_{in}, \rho_{in}$

◊ *Parameters* : η_{∞}, A, V, τ_s, f_s, c_p, r_{tnom}

- *Constants :* R_{shst}
- *Output :* $\rho_{ou}, x_{ou}, w_{ou}$
- *Algebraic equations*:

$$\gamma = \frac{c_p}{c_p - R_{shst}}$$

$$h_o = \frac{x_o}{\rho_o} \qquad *$$

$$T_o = \frac{h_o - h_{in}}{c_p} + T_{in} \qquad *$$

$$p_o = R\,\rho_o\,T_o \qquad *$$

$$m = \frac{\gamma}{\gamma - \eta_\infty(\gamma - 1)}$$

$$r^{(\frac{2}{m})} - r^{(\frac{m+1}{m})} = \frac{w_{ou}^2}{A^2 \rho_o p_o}\left(\frac{m-1}{2\eta_\infty m}\right)$$

Solve for r.

$$p_{ou} = r\,p_o$$

$$T_{ou} = T_o r^{(\eta_\infty \frac{\gamma-1}{\gamma})}$$

$$\Delta h_I = c_p\,T_o\left(r_{hp}^{R/c_p} - 1\right)$$

$$\eta = \frac{1 - r^{(\eta_\infty \frac{\gamma-1}{\gamma})}}{1 - r^{(\frac{\gamma-1}{\gamma})}}$$

$$P = \eta\,\Delta h_I\,w_{ou}$$

$$h_{ou} = h_o + c_p\left(T_{ou} - T_o\right)$$

◊ *Differential equations*:

$$\frac{d}{dt}\left(\rho_{ou}\right)=\frac{1}{V}\left(w_{in}-w_{ou}\right) \qquad *$$

$$\frac{d}{dt}\left(w_{ou}\right)=\frac{1}{\tau_s}\left(w_{in}-w_{ou}\right) \qquad *$$

$$\frac{d}{dt}\left(x_o\right)=\frac{1}{V}\left(w_{in}h_{in}-w_{ou}h_o\right) \qquad *$$

◊ *Outputs* : w_{ou}, ρ_{ou} h_{ou}, P, T_{ou}

The equations marked with * are not applicable to IP section.

<u>High pressure section</u>

For the high pressure section, the following substitutions to the general model must be made:

◊ *Inputs* :

w_{in}, = w_s - from the steam turbine controller

$p_{in} = p_s$ - from the boiler model (superheater)

T_s - from the boiler model (superheater)

$h_s = f_{ST3}(p_s, T_s)$ - [superheated steam table]

$h_{in} = h_s$

◊ *Parameters :*

$\eta_\infty = \eta_{\infty hp}$

$A = A_{hp}$

$V = V_{hp}$

$\tau_s = \tau_{hp}$

$f_s = f_{hp}$

$c_p = c_{php}$

◊ *Outputs :*

$w_{ri} = w_{ouHp} = w_{ou}$

$p_{ri} = p_{ouhp} = p_{ou}$

$h_{ouhp} = h_{ou}$

$P_{hp} = P$

$T_{ri} = T_{ouhp} = T_{ou}$

Intermediate pressure section

For the intermediate pressure section, the following substitutions to the general model must be made:

◊ *Inputs* :

w_{ou}, $= w_{ro}$ - from the boiler model (reheater)

$p_o = p_{ro}$ - from the boiler model (reheater)

$\rho_o = \rho_{ro}$ - from the boiler model (reheater)

T_r - from the boiler model (reheater)

$h_r = f_{STh}(p_{ro}, T_r)$ - [superheated steam table]

$h_{in} = h_r$

◊ *Parameters :*

$\eta_\infty = \eta_{\infty ip}$

$A = A_{ip}$

$c_p = c_{pip}$

◊ *Outputs :*

$w_{inlp} = w_{ouip} = w_{ou}$

$p_{inlp} = p_{ouip} = p_{ou}$

$h_{oulp} = h_{ouip} = h_{ou}$

$P_{ip} = P$

Low pressure section

For the low pressure section, the following substitutions to the general model must be made:

◊ *Inputs* :

$w_{in} = w_{inlp} = w_{ouip}$

$p_{in} = p_{inlp} = W_{ouip}$

$h_{in} = h_{inlp} = h_{ouip}$

◊ *Parameters :*

$\eta_{\infty} = \eta_{\infty lp}$

$A = A_{lp}$

$V = V_{lp}$

$\tau_s = \tau_{lp}$

$f_s = f_{lp}$

$c_p = c_{plp}$

◊ *Outputs :*

$w_{oulp} = w_{ou}$

$p_{oulp} = p_{ou}$

$h_{oulp} = h_{ou}$

$P_{lp} = P$

Power delivered by the turbine

$$P_{mech} = P_{hp} + P_{ip} + P_{lp}$$

4.3.4 Condenser

4.3.4.1 Introduction

A common type of condenser applied in industry is the shell-and-tube type (Fig.4.5). It comprises a bundle of small diameter tubes wholly enclosed within a large diameter pipe or shell. Shell fluid flows over the tube bundle surface through the cavity existing between the shell and the tubes. Tube fluid flows counter-currently and is internal to the tubes. Thus in this device, indirect heat exchange will occur from the warmer fluid to the cooler fluid via the metal tube interface. It is usual for the shell fluid to be the heat source and the tube fluid the heat sink. In the case of the condenser the shell fluid is saturated steam and the tube fluid is cooling water.

A more complicated situation arises if superheated steam enters the heat exchanger and/or partial flooding of the shell-side volume occurs. This would require a model that represented three interaction regions on the shell-side i.e. superheated steam, saturated steam and sub cooled liquid regions.

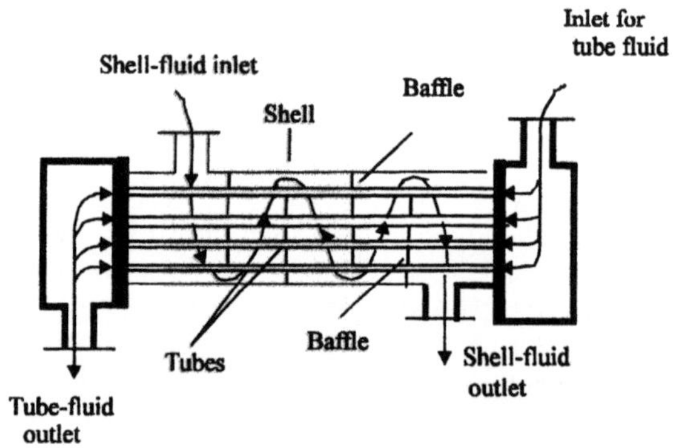

Fig. 4.5 Shell-tube type condenser.

The model to be presented assumes that saturated or near-saturated steam occupies the whole of the shell volume, thus simplifying the analysis. However, the model can be easily extended to represent the more complicated situation discussed above, if this proves necessary in future. The model is capable of predicting the dynamic performance of the heat exchanger over a wide range of operating conditions.

It is noted that the condenser is one of a number of power plant components that are designed to facilitate heat exchange between process fluids. As for the condenser, indirect heat exchange via a metal interface occurs in the boiler, air/water closed heaters etc. By contrast, direct contact heat exchange occurs in the deaerator, see Section 4.3.5.

4.3.4.2 Literature Review

A highly accurate, detailed model of a heat exchanger is presented by [*Ferrarini and Maffezzoni, 1992*]. In their paper the fundamental conservation laws are written with distributed parameters. Lumping the model parameters is avoided since cases where long residence times occur exhibit non-physical oscillations in the time solution. This in turn, can lead to an incorrect prediction of the process delay time, and accurate estimates of the latter are important for controller design. Instead, the characteristic line numerical integration method is used to solve the set of partial differential equations arising from the conservation laws.

Lumped parameter models are reported by [*Kapadia and Wolgemuth, 1984*] and [*Daniel et al., 1984*]. The first of these papers is more complicated in that partial flooding of the condenser is modelled. Other sources include a chapter of the book by Roffel and Rijnsdorp [*1982*] which is devoted to describing a distributed model of a heat exchanger, its reduction to static form and also its linearization.

The preferred model is that described by Daniel et al., [*1984*]. This is due to the clarity of its presentation, the fact that it has been experimentally verified and that it represents the level of complexity desired and can be easily extended to include greater complexity if so desired.

4.3.4.3 Model Assumptions

- Condenser is a shell and tube type.
- Liquid flow tubeside, condensing steam flow shell-side.
- Saturated or near-saturated steam enters the condenser, saturated condensate exits the condenser via a steam-trap. There is no flooding of the condenser.

- Thermal masses of condenser material and fluids are lumped as follows:
  - shell material = one node
  - shell tubes - one node
  - shell steam = one node
  - tube liquid = N nodes existing along the condenser axis.
- Shell is perfectly insulated.
- Heat transfer is radial only, axial heat conduction is ignored.
- Flow is one-dimensional.
- There is no evaporation of the tube liquid.
- Time delay of liquid in the condenser ignored.
- Liquid is ideally mixed over the tube cross-sectional area.
- Heat transfer coefficients are constants.

4.3.4.4 Condenser Equations

<u>Shell material equations</u>

Heat flow to shell from steam

$$Q_{sh} = H_{sh}A_{sh}(T_{st}-T_{sh})$$

where :

Q_{sh} -	*heat flow to shell*
H_{sh} -	*heat transfer coefficient for shell material*
A_{sh} -	*surface area of the shell*
T_{st} -	*steam temperature*
T_{sh} -	*shell temperature*

Heat capacitance of shell

$$C_{sh} = m_{sh}\, c_{psh}$$

where :

C_{sh} -	*heat capacitance of shell*
m_{sh} -	*mass of shell*

c_{psh} - *specific heat of shell material*

Temperature dynamics of the shell

$$\frac{d}{dt}T_{sh} = Q_{sh}\frac{1}{C_{sh}}$$

<u>Tubes material equations</u>

Thermal capacitance of tubes material

$$C_t = N_t m_t c_{pt}$$

where :

C_t - *thermal capacitance of tubes*

N_t - *number of tubes*

m_t - *mass of one tube*

c_{pt} - *specific heat of tube*

Temperature dynamics

$$\frac{d}{dt}T_t = \frac{1}{C_t}(Q_{st} - Q_{hf})$$

where :

T_t - *tubes temperatures*

Q_{st} - *heat flow from steam to tubes*

Q_{hf} - *heat flow from tubes to the cooling liquid*

Heat flow from steam to tubes

$Q_{st} = (T_{st} - T_t)/\varphi_t$

$$\varphi_t = \frac{L_n(r_o / r_i)}{2(2\pi k_t L_t N_t)} + \frac{1}{N_t A_{ot} H_{st}}$$

where :

r_o - *outside radius of tubes*

r_i - *inside radius of tubes*

k_t - *thermoconductivity of tube walls*

L_t - *length of tubes*

A_{ot} - *surface area of tubes*

H_{st} - *heat transfer coefficient for tubes material (steam to tubes)*

Heat flow from tubes to the cooling liquid

$$Q_{hf} = \sum_{j=1}^{N} Q_{hfi}$$

where :

N - number of sections in the heat exchanger

$$Q_{hfi} = \frac{T_t - \overline{T}_{fi}}{\varphi_{ti}}$$

Q_{hfi} - *heat flow in one section*

$\overline{T}_{fi}$ - *average liquid temperature in the section*

φ_{ti} - *is defined in the same way as φ_t except that L_t must be substituted with ΔL_{ti} the section length and A_{ot} must be substituted with ΔA_{ti} the surface area of tubes in the section*

$$\Delta L_{ti} = \frac{1}{N} L_t$$

$$\Delta A_{ti} = \frac{1}{N} A_{ot}$$

Steam equations

Temperature dynamics

$$\frac{d}{dt} T_{st} = \frac{w_{st}(h_{st} - h_{cond}) - Q_{sh} - Q_{st}}{V\left[\rho_{st}\left(\frac{\partial h}{\partial T}\right)_{sat} + \left(h_{st} - h_{cond}\right)\left(\frac{\partial \rho}{\partial T}\right)_{sat} - \left(\frac{\partial P}{\partial T}\right)_{sat}\right]}$$

where :

w_{st} - *mass flow of steam*

h_{st} - *specific enthalpy of saturated steam*

h_{cond} - *specific enthalpy of condensate (saturated liquid)*

V - *volume of steam in the shell (constant)*

ρ_{st} - *density of saturated steam*

$\left(\frac{\partial h}{\partial T}\right)_{sat}$ - *partial derivative of saturated steam enthalpy with temperature at given T_{st} (from saturated steam tables).*

$\left(\frac{\partial \rho}{\partial T}\right)_{sat}$ - *partial derivative of saturated steam density with temperature at given T_{st} (from saturated steam tables).*

$\left(\frac{\partial p}{\partial T}\right)_{sat}$ - *partial derivative of saturated steam pressure with temperature at given T_{st} (from saturated steam tables).*

Note that:

$$V\left[\rho_{st}\left(\frac{\partial h}{\partial T}\right)_{sat} + \left(h_{st} - h_{cond}\right)\left(\frac{\partial \rho}{\partial T}\right)_{sat} - \left(\frac{\partial p}{\partial T}\right)_{sat}\right] = C_{st}$$

where :

C_{st} - *thermal capacitance of steam in the shell*

Condensate pressure

$$p_{cond} = f(T_{st})$$

where :

p_{cond} - *pressure of condensate at saturation conditions (from steam tables)*

Condensate mass flow

$$w_{cond} = w_{st} - \frac{d}{dt} T_{st}\left(\frac{\partial \rho}{\partial T}\right)_{sat}$$

where :

w_{cond} - *condensate mass flow.*

Liquid equations

Thermal capacitance of j-th section

$$C_{fj} = \rho_f C_{pf} V_{fj}$$

where :

C_{fj} - *thermal capacitance of liquid in a j-th section*

ρ_f - *density of liquid*

C_{pf} - *specific heat at constant pressure*

V_{fj} - *volume of liquid in j-th section*

Temperature dynamics of j-th section

$$\frac{d}{dt}T_{fj} = \frac{Q_{hfj} + w_f C_{pfj}(T_{fj-1} - T_{fj})}{C_{fj}}$$

where :

T_{fj} - *temperature at the end of j-th section*

Q_{hfj} - *heat flow from tubes to cooling liquid in j-th section (as defined earlier)*

Liquid pressure drop

$$\Delta p_f = \sum_{j=1}^{N} \Delta p_{fj}$$

$$\Delta p_{fj} = f_f \frac{w_f^2}{\rho_f}$$

where :

Δp_f - *overall liquid pressure drop*

Δp_{fj} - *liquid pressure drop in j-th section*

f_f - *friction coefficient*

w_f - *liquid mass flow*

ρ_f - *liquid density*

4.3.4.5 Block diagram

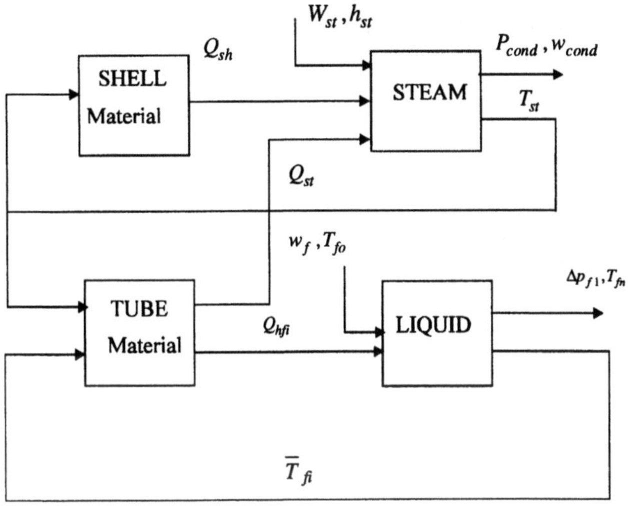

Fig. 4.6 Condenser - block diagram

4.3.4.6 Model structure for implementation as an algorithm

Shell material equations

- *Inputs* : T_{st}
- *Parameters* : $H_{sh}, A_{sh}, m_{sh}, c_{psh}$
- *States* : T_{sh}
- *Algebraic equations :*

$$Q_{sh} = H_{sh}\, A_{sh}(T_{st}-T_{sh})$$

- *Differential equations*

$$\frac{d}{dt}T_{sh} = \frac{1}{m_{sh}c_{psh}}Q_{sh}$$

- *Outputs :* Q_{sh}

Tube material equations

- *Inputs :* T_{st}, $\overline{T}_{fi}$ (i=1,..,N)
- *Parameters :* r_o, r_i, k_t, L_t, N_t, A_{ot}, H_{st}, N, H_{tl}
- *States :* T_t
- *Algebraic equations:*

$$\varphi_t = \frac{\log\left(\frac{r_o}{r_i}\right)}{2(2\pi k_t L_t N_t)} + \frac{1}{N_t A_{ot} H_{st}}$$

$$Q_{st} = \frac{1}{\varphi_t}\left(T_{st} - T_t\right)$$

$$\varphi_{ti} = \frac{\log\left(\frac{r_o}{r_i}\right)}{2(2\pi k_t L_t \frac{N_t}{N})} + \frac{1}{\frac{N_t}{N} A_{ot} H_{tl}}$$

$$Q_{hfi} = \frac{1}{\varphi_{ti}}\left(T_t - \overline{T}_{fi}\right) \qquad i = 1,\cdots,N$$

$$Q_{hf} = \sum_{j=1}^{N} Q_{hfi}$$

- *Differential equation:*

$$\frac{d}{dt}T_t = \frac{1}{N_t m_t c_{pt}}(Q_{st} - Q_{hf})$$

- *Outputs:* Q_{st}, Q_{hfi} (i=1 ,.., N)

Steam equations

- *Inputs :* w_{st}, h_{st}, Q_{sh}, Q_{st}
- *Parameters :* V
- *States :* T_{st}
- *Algebraic equations :*

$$\left(\frac{\partial h}{\partial T}\right)_{sat} = f_{He1}(T_{st}) \qquad \text{[saturated steam table]}$$

$$\left(\frac{\partial p}{\partial T}\right)_{sat} = f_{He2}(T_{st})$$ [saturated steam table]

$$\left(\frac{\partial p}{\partial T}\right)_{sat} = f_{He3}(T_{st})$$ [saturated steam table]

$$h_{cond} = f_{He4}(T_{st})$$ [saturated steam table]

$$\rho_{st} = f_{He5}(T_{st})$$ [saturated steam table]

$$p_{cond} = f_{He6}(T_{st})$$ [saturated steam table]

$$C_{st} = V\left[\rho_{st}\left(\frac{\partial h}{\partial T}\right)_{sat} + \left(h_{st} - h_{cond}\right)\left(\frac{\partial p}{\partial T}\right)_{sat} - \left(\frac{\partial p}{\partial T}\right)_{sat}\right]$$

$$d_{Tst} = \frac{1}{C_{st}}\left[w_{st}\left(h_{st} - h_{cond}\right) - Q_{sh} - Q_{st}\right]$$

$$w_{cond} = w_{st} - d_{Tst}\left(\frac{\partial \rho}{\partial T}\right)_{sat}$$

◊ *Differential equations:*

$$\frac{d}{dt}T_{st} = d_{Tst}$$

◊ *Outputs :* Tst, pcond, wcond

Liquid equations

◊ *Inputs :* w_f, T_{fo}, Q_{hfi} (=1,...,N)

◊ *Parameters :* N, ρ_f, c_{pf}, V_{fi}, f_f

◊ *States :* T_{fi} (i=1,..,N)

Equations for i-th section (i=1,..,N)

◊ *Algebraic equations*

$$C_{fi} = \rho_f, c_{pf} V_{fi}$$

$$\Delta p_{fi} = f_f \frac{w_f^2}{\rho_f}$$

$$\overline{T}_{fi} = \frac{1}{2}\left(T_{fi-1} + T_{fi}\right) i = 1,..,N$$

◊ *Differential equations:*

$$\frac{d}{dt}T_{fi} = \frac{1}{C_{fi}}\left[Q_{hfi} + w_f c_{pf}(T_{fi-1} - T_{fi})\right]$$

◊ *Equation for the liquid total pressure drop*

$$\Delta p_f = \sum_{i=1}^{N} \Delta p_{fi}$$

◊ *Outputs:* $\Delta p_f, T_{fN}, \overline{T}_{fi}\,(i = 1,..,N)$

4.3.5 Feedwater System

4.3.5.1 Introduction

By appropriately connecting the following components the feedwater system may be modelled :

- deaerator
- pumps
- valves
- actuators
- pipes

It is noted that potential energy savings exist in the adoption of a supervisory control strategy that regulates the feedwater pump speed set point as a function of attemporator spray flow measurement. Operation of the feedwater pump consumes approximately 2% of the power output of a typical steam turbine power plant. The description of the above sub-modules now follows:

4.3.5.2 The Economiser

The economiser equations have exactly the same structure as the boiler reheater equations, see section 4.3.1.

Model Assumptions

(i) The economiser variables are lumped, ie no spatial dependency only time dependency.

(ii) The thermodynamic properties of non-saturated water are approximated by saturation relationships.

It is expected that neglecting the spatially distributed variations in liquid temperature in the economiser will cause a significantly greater modelling inaccuracy than is the case for the reheater. This is because of the relatively large thermal capacitance of water, whereas steam possesses a low thermal capacitance resulting in small spatial temperature variations. Improved model accuracy for the economiser can easily be incorporated by dividing it into discrete sections, as is done for the tube-side equations in the condenser model, see Section 4.3.4

Economiser Equations

Mass balance

$$w_{ei} - w_{eo} = V_e \frac{d}{dt}(\rho_e)$$

where :

w_{ei} - *economiser inlet feedwater flow*

w_{eo} - *economiser outlet water flow*

V_e - *economiser volume*

ρ_e - *density of liquid internal to economiser*

Gas-tube heat balance

$$Q_{es} = Q_e + M_e C_e \frac{d}{dt}(T_{et})$$

where :

Q_{es}- *heat transfer by convention from furnace, absorbed by the economiser surface.*

Q_e - *heat transferred to the economiser liquid*

M_e - *mass of economiser tubes*

C_e - *heat capacitance of economiser tubes*

T_{et} - *temperature of economiser tubes*

Heat transfer tube-liquid

$$Q_e = k_e w_{ei}^{0.8}(T_{et} - T_{eo})$$

where :

k_e - *an empirical cofficient*

T_{eo} - *economiser liquid temperature*

Friction loss (momentum balance)

$$p_{ei} - p_{eo} = \frac{w_{ei}^2}{\rho_e} f_e$$

where:

p_{ei} - *economiser inlet water pressure*

p_{eo} - *economiser outlet water pressure*

f_e - *a friction coefficient*

Heat balance for liquid

$$Q_e + w_{ei}h_{ei} = w_{eo}h_{eo} + V_e \frac{d}{dt}(\rho_e h_{eo})$$

where :

h_{ei} - *inlet water specific enthalpy*

h_{eo} - *outlet water specific enthalpy*

Inputs to the economiser model :

p_{ei}- *inlet water pressure (from the deaerator)*

T_{ei} - *inlet water temperature (from the deaerator)*

Q_{es}- *heat radiation absorbed by the economiser*

w_{eo}- *outlet water mass flow*

$w_{eo} = w_d + w_a$, where w_d = *downcomer liquid mass flow (from the boiler drum + downcomer module) and* w_a *is attemperator liquid mass flow*

The outputs from the economiser which are internal to the boiler model are:

T_{et} - *economiser metal tube temperature*

p_{eo} - *economiser outlet water pressure*

T_{eo} - *economiser outlet water temperature*

Model Structure for Implementation as an Algorithm

- *Inputs:* $p_{ei}, w_{ei}, w_{eo}, T_{ei}, Q_{es}$
- *Parameters* : f_e, k_e, V_e, M_e, c_e
- *States* : ρ_e, T_{et}, X_{e1}
- *Algebraic equations:*

$$h_{eo} = \frac{X_{e1}}{\rho_e}$$

$$h_{ei} = f_{16}(p_{ei})$$ [approximation using sat. liquid tables]

$$T_{eo} = f_{17}(h_{eo})$$ [approximation using sat. liquid tables]

$$p_{eo} = f_{18}(h_{eo})$$ [approximation using sat. liquid tables]

$$Q_e = k_e w_{ei}^{0.8}(T_{et} - T_{eo})$$

- *Differential equations* :

$$\frac{d}{dt}\rho_e = \frac{1}{V_e}(w_{ei} - w_{eo})$$

$$\frac{d}{dt}T_{et} = \frac{1}{Mc_{e_e}}(Q_{es} - Q_e)$$

$$\frac{d}{dt}X_{e1} = \frac{1}{V_e}(Q_e + w_{ei}h_{ei} - w_{eo}h_{eo})$$

- *Outputs*: $T_{et}, p_{eo}, T_{eo}, \rho_e, h_{eo}$

4.3.5.3 Deaerator

A deaerator performs three primary functions as follows:

- feedwater heating
- feedwater de-oxygenation
- feedwater storage

The feedwater is heated to saturation conditions by bringing it into direct contact with the extracted superheated steam from the steam turbine. This contrasts with the indirect heat exchange that occurs in the boiler and condenser components, where a metal interface exists between the two process fluids.

Two types of deaerator predominate in industrial power plants. These are the vacuum deaerator and pressure deaerator respectively. Vacuum deaerators are operated below atmospheric pressure at a saturation temperature of 82°C and are capable of reducing the oxygen content in the water to less than 0.02 mg per litre [*Gunn and Horton, 1989*]. Extra cost results from maintaining the vacuum conditions which requires the application of vacuum pumps or steam ejectors. Pressure deaerators are operated above atmospheric pressure and can reduce the oxygen content to 0.02 mg per litre. The pressure inside the deaerator is controlled by adjusting the extraction steam supply to the vessel. Pressure regulation also constrains the saturation temperature to be greater than 105°C.

At the saturation condition obtained inside the deaerator vessel, the solubility of gas in the feedwater is very low. Thus any non-condensable gases are separated from the feedwater and are subsequently expelled to the atmosphere through a vent.

The storage of a volume of feedwater inside the deaerator provides spare capacity in meeting future boiler feedwater demands that varies in accordance with varying power plant load conditions.

The superheated extraction steam releases heat energy to the feedwater. Equilibrium is obtained when both the steam and water reach saturation conditions inside the deaerator. Then no further heat exchange occurs between the two phases and there is no evaporation or condensation inside the vessel.

<u>Model Assumptions</u>

The main assumptions employed in constructing the model are as follows:

(i) The steam phase is not modelled and all the steam turns to water.

(ii) The de-oxygenation process is not modelled.

(iii) There are no temperature gradients in the deaerator liquid, due to turbulence effects (Perfect mixing assumed).

(iv) The deaerator steam and liquid phases are always at or near to the deaerator's rated saturate operating condition. This enables linearisation of the evaporation/ condensation dynamics about the rated operating point.

(v) The feedwater outlet pipe is situated at the base of the deaerator.

(vi) The entrained water remains at or near to saturated condition due to the action of feedback control.

Deaerator equations

Liquid mass balance:

$$w_c + w_m - w_f - w_{ec} = \frac{d}{dt}(m_f)$$

where:

w_c - *mass flow of condensate*

w_m - *mass flow of makeup water*

w_f - *mass flow of outlet feedwater*

w_{ec} - *mass flow of evaporating liquid or condensing steam*

m_f - *mass flow of liquid in deaerator*

Liquid heat balance

$$w_c h_c + w_m h_m - w_f h_f - w_{ec} h_{eL} = \frac{d}{dt}(m_f h_f)$$

where :

h_c - *specific enthalpy of condensate*

h_m - *specific enthalpy of makeup water*

h_f - *specific enthalpy of outlet feedwater*

h_{eL} - *specific enthalpy of saturated liquid*

Liquid volume

$$V_f = \frac{m_f}{\rho_f}$$

where:

V_f - *volume of feedwater in deaerator*

ρ_f - *density of feedwater in deaerator*

Liquid level

$$V_f = f(L) \Rightarrow L = f^{-1}(V_f)$$

where:

L - *deaerator liquid level*

f(.) - *a function dependent upon the shape of the deaerator vessel*

Outlet feedwater pressure

$$p_{fo} = p_f + \rho_f gL$$

where :

p_{fo} - *outlet feedwater pressure*

p_f - *pressure of water inside the deaerator*

g - *gravitational constant*

Deaerator algorithmic implementation

- *Inputs* : $w_f, w_c, w_m, w_x, h_c, h_m, h_x, p_s$
- *Parameters*: V_{dae}, r
- *States* : m_f, x_{01}
- *Algebraic equations:*

$$h_f = X_{01} / m_f$$

$$\rho_f = f_1(h_f) \quad \text{[saturated water tables]}$$

$$T_f = f2(h_f) \quad \text{[saturated water tables]}$$

$$p_f = f_7(h_f) \quad \text{[saturated water tables]}$$

$$V_f = m_f / \rho_f$$

$$L = f^{-1}(V_f) \quad \text{[f(.) : function dependent on deaerator shape]}$$

$$p_{fo} = p_s + \rho_f gL$$

- *Differential equations:*

$$\frac{d}{dt} m_f = w_c + w_m - w_f + w_x$$

$$\frac{d}{dt} X_{01} = w_c h_i + w_m h_m - w_f h_f + w_x h_x$$

◊ *Outputs:* p_{fo}, L

4.3.5.4 Pump

The function of a pump is to transfer a fluid from one location to another. Fluid flow is achieved despite restrictions in the flow circuit by increasing the pressure of the fluid in the pump. Centrifugal pumps are commonly used in power plant applications. They are nearly always modelled by their steady state pressure-flow characteristic [*Maffezzoni,* 1992]. This type of model will be presented here.

Literature Review

Various textbooks describe pump performance under steady-state conditions. See, for example, [*Shinskey, 1979*], [*Sorensen, 1983*] and [*Roffel and Rijnsdorp, 1982*]. The preferred presentation applied in the development of the pump model is that given in [*Shinskey, 1979*], with the addition of an actuator for speed change under closed loop control.

Model Assumptions

(i) dynamics ignored

(ii) incompressible fluid flow

(iii) flow is controlled via speed variation. (This is shown to be more efficient than throttling see [*Roffel and Rijnsdorp, 1983*]).

(iv) Constant load (= resistance to flow) presented by apparatus, valves and pipes existing in the hydraulic circuit.

(v) centrifugal pump type

(vi) constant pump efficiency

Pump equations

The development of [*Shinskey, 1979*] is followed here.

Pump head as a function of speed and flow

$$\frac{\Delta p}{\rho} = K_1 N^2 - K_2 F^2$$

where:

Δp - *pressure rise across the pump*

$\overline{\rho}$ - *fluid density reference value*

N - *pump speed*

F - *volumetric flow*

K_1, K_2 - *machine construction constants*

Load lines are represented by

$$\frac{\Delta p_q}{\overline{\rho}} = K_3 F^2$$

where :

Δp_q - *pressure drop across load*

K_3 - *constant dependent on the load*

Pump pressure rise

$$\Delta p = \Delta p_s + \Delta p_q$$

where :

ΔP_s - *static pressure rise across the pump*

Combining the above equations leads to an equation for flow at constant load:

$$F = \sqrt{\frac{K_1 N^2 - \Delta p_s / \overline{\rho}}{K_2 + K_3}}$$

Power required for pumping the fluid

$$P_w = \frac{\Delta p F}{\eta} \times \frac{100}{1}$$

where :

P_w - *pump power consumption*

η - *pump efficiency*

Speed change dynamics

see model for actuator.

Model structure for implementation as an algorithm

- ◊ *Inputs* : N_{dem}, ρ, p_i
- ◊ *Parameters*: $k_1, k_2, k_3, \eta, \bar{\rho}, \Delta p_s$
- ◊ *Algebraic equations*:

$$F = \sqrt{\frac{k_1 N^2 - \Delta p_s / \bar{\rho}}{k_2 + k_3}}$$

$$\Delta p_q = k_3 \bar{\rho} F^2$$

$$\Delta p = \Delta p_s + \Delta p_q$$

$$p_o = p_i + \Delta p$$

$$w = F\rho$$

$$P_w = \frac{\Delta p F}{\eta} x \frac{100}{1}$$

Outputs: w, P_w, ρ, p_o

4.3.5.5 Valve

Valves are used extensively in power plant to control the mass flow rate of fluids via the adjustment of flow area through an orifice.

Literature Review

A number of textbooks provide an introduction to valve operation and performance. Two examples are [*Balchen and Mumme, 1988*] and [*Stephanopoulos, 1984*]. Valve flow characteristic and plug designs are shown in [*Stephanopoulos, 1984*]. A more detailed steady-state analysis of valves is provided by [*Knowles, 1990*].

Model Assumptions

(i) Pressure drop across valve is constant

(ii) gas flow is uniform polytropic

(iii) ratio of specific heats for gas is constant

Model equations

[*Balchen and Mumme, 1988*] present flow equations similar to the following:

$$w_l = c_{v_1} f(x)\sqrt{\rho_1 \Delta p} \qquad \text{LIQUID FLOW}$$

$$w_g = c'_{v_2} f(x)\sqrt{\rho_{ig} \Delta p} \qquad \text{GAS FLOW}$$

where:

w_L - *mass flow of liquid*

w_g - *mass flow of gas*

$f(x)$ - *valve characteristic*

x - *valve stem position*

ρ_L - *liquid density*

c_{v1} - *constant valve construction coefficient*

c_{V2} - *constant coefficient*

$$c'_{v2} = c_{v2} * (1 - \frac{0.466}{k}\frac{\Delta p}{\Delta p_{rated}})$$

where :

c_{v2} - *constant valve construction coefficient*

k - *ratio of specified heats for gas*

Δp_{rated} - *rated pressure drop across valve*

ρ_{ig} - *gas inlet density*

Δp - *pressure drop across valve (constant)*

In addition, for gas flow under the assumption of a uniform polytropic expansion through the valve, the following relation holds:

$$\rho_{og} = \rho_{ig}\left(\frac{p_o}{p_i}\right)^{-\frac{1}{k}}$$

where:

ρ_{og} - *outlet gas density*

p_o - *outlet pressure (=* p_i *-* Δp*)*

$$\frac{T_{og}}{T_{ig}} = \left(\frac{p_{og}}{p_{ig}}\right)^{(k-1)/k}$$

where:

T_{og} - *outlet gas temperature*

T_{ig} - *inlet gas temperature*

$$h_{og} = h_{ig} + c_p\left(T_{og} - T_{ig}\right)$$

where:

h_{og} - *outlet gas enthalpy*

h_{ig} - *inlet gas enthalpy*

c_p - *specific heat at constant pressure*

Block Diagram

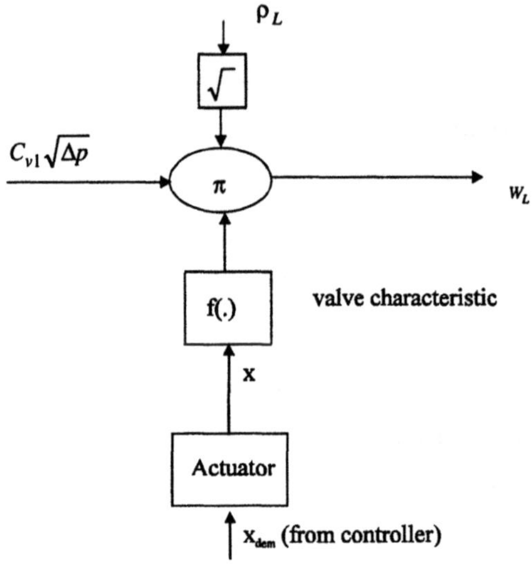

Fig. 4.7 Valve liquid - block diagram.

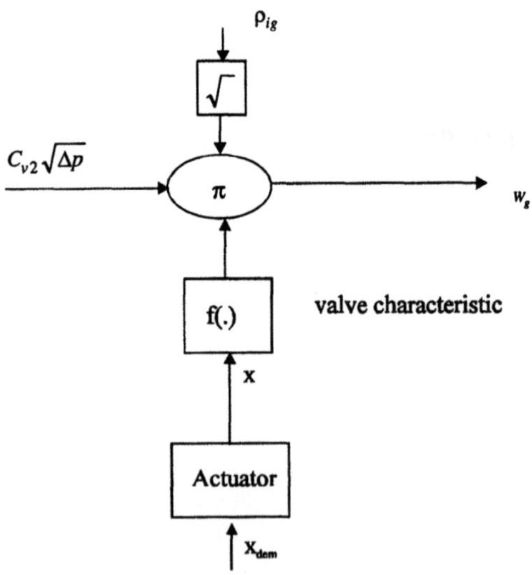

Fig. 4.8 Valve gas - block diagram.

Model structure for implementation as an algorithm

LIQUID FLOW

- *Inputs* : x, ρ_l, p_i
- *Parameters*: $c_{v1}, \Delta p$
- *Known functions* : f(.)
- *Algebraic equations*:

$$w_L = c_{v1} f(x) \sqrt{\rho_L \Delta P}$$

$$p_o = p_i - \Delta p$$

- *Outputs*: w_L, ρ_L, p_o

GAS FLOW

- *Inputs* : $x, \rho_{ig}, P_{ig}, T_{ig}, h_{ig}$
- *Parameters* : $c_{v2}, k, \Delta p, \Delta p_{rated}, c_p$
- *Known functions* : f(.)
- *Algebraic equations*:

$$c'_{v2} = c_{v2}\left(1 - \frac{0.466}{k}\frac{\Delta p}{\Delta p_{rated}}\right)$$

$$w_g = c'_{v2} f(x) \sqrt{\rho_{ig} \Delta p}$$

$$p_o = p_i - \Delta p$$

$$\rho_{og} = \rho_{ig}\left(\frac{p_o}{p_i}\right)^{-1/k}$$

$$\frac{T_{og}}{T_{ig}} = \left(\frac{p_{og}}{p_{ig}}\right)^{(k-1)/k}$$

$$h_{og} = h_{ig} + c_p\left(T_{og} - T_{ig}\right)$$

- *Outputs*: $w_g, p_o, \rho_{og}, h_{og}, T_{og}$

4.3.5.6 Actuator

The actuator model includes rate and position limiting, and is shown in Fig. 4.9.

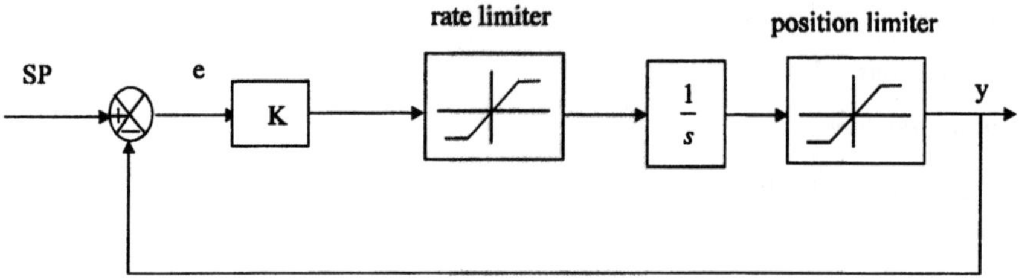

Fig. 4.9 Actuator - block diagram.

4.3.5.7 Feedwater system - block diagram.

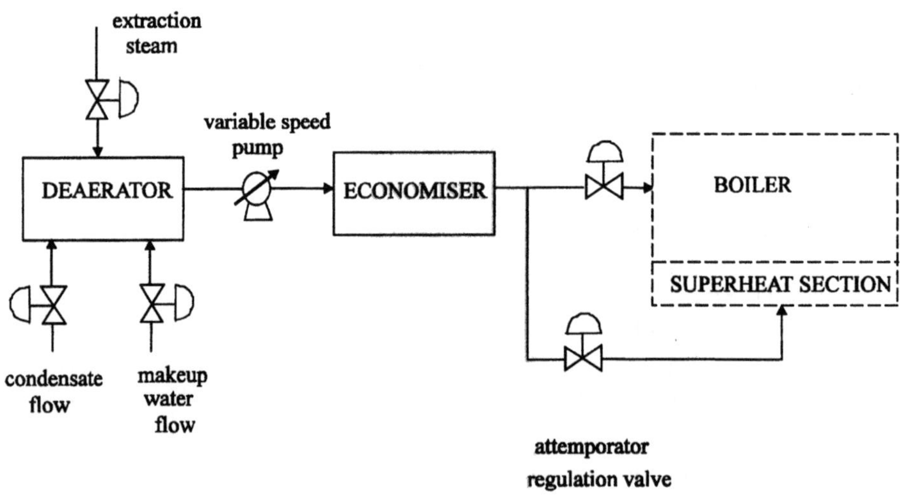

Fig. 4.10 Feedwater system - block diagram.

4.3.6 Electrical Generator

4.3.6.1 Introduction

In the synchronous machine applied as a generator, current is supplied to the rotor field winding. The resulting magnetic flux cuts the stator windings, and the construction is such that three phase sinusoidal electrical currents circulate in the stator windings under load conditions. These currents in turn, produce a rotating magnetic field whose angular speed equals that of the rotor. An electro-magnetic torque is set up that opposes the mechanical drive torque of the prime-mover. Under steady-state conditions the prime-mover torque is exactly balanced by the electrical load torque, assuming no losses.

Two generator models are presented in this section. The first represents the generator as a voltage behind an impedance, and predicts the voltages, currents and real/reactive power variations under transient conditions. The second is far simpler, only including real power and frequency variations and neglecting voltages,currents and reactive power. It is only valid for relatively small load disturbances, corresponding to those normally experienced when connected to a large power generation grid.

The particular model to be used depends upon the proposed application. If the objective is to interface to an electrical network model then the first model provides the necessary complexity. If on the other hand the research objectives are concerned only with the process side of a power generation plant , then it may be appropriate to avoid the modelling of the electrical behaviour of the generator altogether. Instead electrical load variations are modelled as a stochastic input disturbance to the equations that govern the mechanical behaviour of the machine.

4.3.6.2 Literature Review

Models that represent the generator as a voltage behind an impedance, defining model variables in an axes frame that rotates synchronously with the machine, are common, see, for example [*Arrillaga & Arnold, 1990*], [*Rafian et al., 1987*].

The simplified generator model in block diagram form is described by [*Wood and Wollenberg, 1984*]. Load variations are modelled as a stochastic processs in [*de Mello et al.,*

1973], whilst typical time histories of grid frequency variations (which are dependent on load variations) are presented by [*Cegrell, 1986*].

A simple power system consisting of a number of connected generators within a local generation area, tied to an infinite capacity external area, is described by [*de Mello et al., 1973*]. This permits the simulation of synchronising power oscillations that typically occur between connected machines as a result of load disturbances. The paper includes a highly simplified model of the electrical network. In [*Rafian et al., 1987*], the network model is more fully developed to include representations of electrical generators, transmission lines, transformers, static compensators, Automatic Voltage Regulators (AVR) and loads.

4.3.6.3 Model Assumptions

MODEL 1 VOLTAGE BEHIND AN IMPEDANCE

- **(i)** The generator may be represented by a voltage behind an impedance.
- **(ii)** The electrical network is represented by an infinite bus.
- **(iii)** Generator parameters are independent of frequency.
- **(iv)** No magnetic saturation occurs.
- **(v)** Steady state and transient behaviour modelled. Very fast (100th's of seconds) so called sub-transient behaviour not modelled.
- **(vi)** Transient voltage fluctuations due to damper windings in the machine direct axis are neglected.
- **(vii)** Armature resistance negligible.
- **(viii)** No hysteresis losses in the iron.
- **(ix)** Rotor speed does not vary greatly from synchronous speed.

MODEL 2 MECHANICAL WITH ELECTRICAL LOAD AS STOCHASTIC INPUT DISTURBANCE

- **(i)** Rotor speed does not vary greatly from synchronous speed.
- **(ii)** Windage and friction losses due to rotation are ignored.

4.3.6.4 Generator Equations

Model 1

The generator is modelled as a voltage behind an impedance. The electrical grid is assumed to have an infinite generation capacity (infinite bus) which is modelled as a constant source of voltage and frequency. Fig 4.11 shows the circuit diagram of the generator connected to the infinite bus.

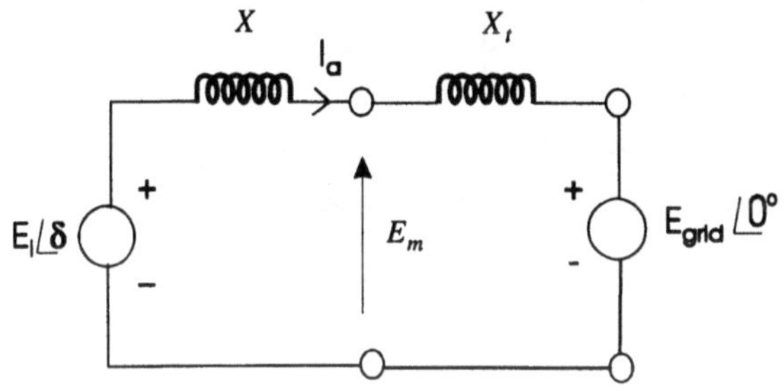

Fig. 4.11 Generator connected to the infinite bus. Key : Ei - *excitation voltage*, Em - *machine terminal voltage*, Egrid - *grid voltage*, X - *machine reactance*, Xt - *transmission reactance*, Ia - *stator (armature) current*, δ - *power angle*.

All model variables are defined in a synchronously rotating axes frame which is fixed with respect to the machine rotor. Fig. 4.12 shows the phasor diagrams, all variables represented in the diagrams are defined below.

Power angle

$$\dot{\delta} = \omega_r - \omega_{grid}$$

where :

ω_r *rotor speed*

ω_{grid} *grid frequency*

Electrical torque

$$T_e = \frac{P_e P_{base}}{\omega_r}$$

i) transient state

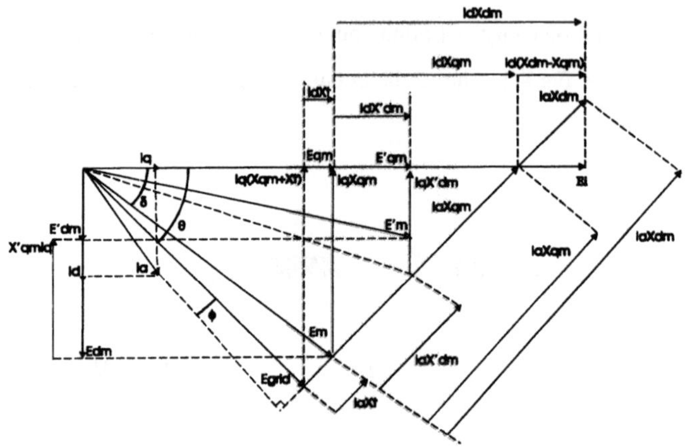

ii) steady state

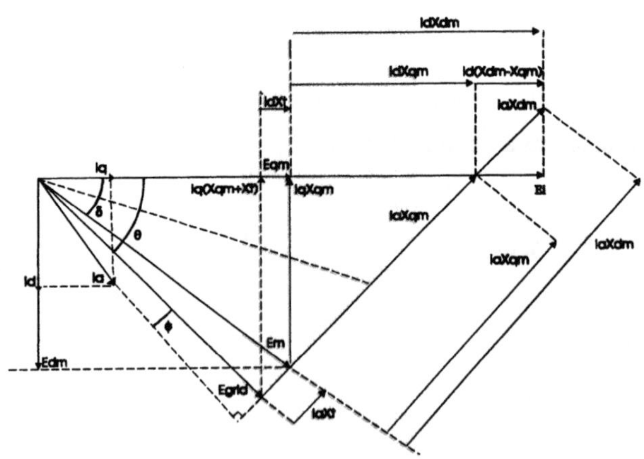

Fig. 4.12 Generator phasor diagrams.

where :

T_e *electrical torque*

P_e *real electrical power*

P_{base} *machine base electrical power*

Real power

$$P_e = \left[E_i - I_d\left(x_{dm} - x_{qm}\right)\right]I_q$$

where :

I_d *stator d-axis current*

I_q *stator q-axis current*

x_{dm} *d-axis reactance*

x_{qm} *q-axis reactance*

Transient q axis voltage

$$\dot{E}'_{qm} = [\left(x'_{dm} - x_{dm}\right)I_d + E_f - E'_{qm}]\tau_{d0}$$

where :

E'_{qm} *transient q-axis voltage*

E_f *field winding voltage*

τ_{d0} *empirical time constant*

Excitation voltage

$$E_i = \left(x_{dm} - x'_{dm}\right)I_d + E'_{qm}$$

q-axis current

$$I_q = \frac{\left(\dfrac{E_{grid}}{E_{base}}\sin\delta\right)}{\left(x_{qm} + x_t\right)}$$

where :

E_{base} *machine base voltage*

q-axis voltage

$$E_{qm} = (E_i - I_d x_{dm})\frac{\omega_r}{\omega_{grid}}$$

d-axis current

$$I_d = \frac{\left(E_i - \frac{E_{grid}}{E_{base}}\cos\delta\right)}{(x_{dm} + x_t)}$$

d-axis voltage

$$E_{dm} = I_q x_{qm}$$

Machine terminal voltage

$$E_m = \sqrt{E_{dm}^2 + E_{qm}^2}$$

Rotor angle

$$\theta = \sin^{-1}\left(\frac{E_{dm}}{E_m}\right)$$

Field current

$$I_f = \left(\frac{E_f}{R_f}\right)E_{fbase}$$

where :

I_f *field current*

R_f *field coil resistance*

E_{fbase} *field base voltage*

Stator current

$$I_a = \sqrt{I_q^2 + I_d^2}$$

Reactive power

$$P_Q = E_{qm} I_d - E_{dm} I_q$$

Shaft speed (swing equation)

$$\dot{\omega}_r = \frac{\left[T_m - T_e - (k_d + k_e)\omega_r\right]}{I}$$

where :

T_m *mechanical drive torque*

k_d *frictional plus windage damping factor*

k_e *electrical damping factor*

I *machine inertia*

Model 2

This model only includes the swing equation above, plus a stochastic noise generator to represent the real power electrical load fluctuations. The swing equation is written in a slightly different form in terms of power rather than torques.

$$\frac{d\omega}{dt} = \frac{\omega_0}{2H_{eq}}\left[P_{mech} - P_{el} - D_{eq}(\omega - \omega_0)\right]$$

where :

ω *frequency of the generator*

H_{eq} *inertia constant for machine*

P_{mech} *mechanical shaft power*

P_{el} *real electrical load power*

D_{eq} *electrical damping coefficient*

ω_0 *synchronous frequency*

4.3.6.5 Block diagram

Fig. 4.13 shows the block diagram of the model.

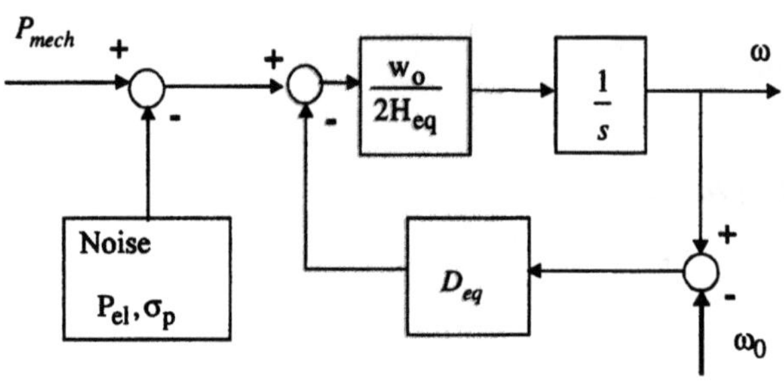

Fig. 4.13 Generator - block diagram of Model 2

4.3.6.6 Algorithm Implementation

Model 1

- *Inputs:* $E_{grid}, \omega_{grid}, P_m, E_f$
- *Parameters:* $P_{base}, E_{base}, x_{dm}, x_{qm}, x_t, x'_{dm}, E_{fbase}, R_f$
- *States:* $\delta, E'_{qm}, \omega_r$
- *Algebraic equations :*

$$T_m = \frac{P_m}{\omega_r}$$

$$\delta = \mathrm{mod}(\delta, 2\pi)$$

$$I_q = \frac{\left(\frac{E_{grid}}{E_{base}} \sin\delta\right)}{\left(x_{qm} + x_t\right)}$$

$$I_d = \frac{\left(E'_{qm} - \frac{E_{grid}}{E_{base}}\cos\delta\right)}{\left(x_t + x'_{dm}\right)}$$

$$E_i = \left(x_{dm} - x'_{dm}\right)I_d + E'_{qm}$$

$$E_{qm} = \left(E_i - I_d x_{dm}\right)\frac{\omega_r}{\omega_{grid}}$$

$$E_{dm} = I_q x_{qm}$$

$$E_m = \sqrt{E_{dm}^2 + E_{qm}^2}$$

$$P_e = \left[E_i - I_d\left(x_{dm} - x_{qm}\right)\right]I_q$$

$$T_e = \frac{P_e P_{base}}{\omega_r}$$

$$\theta = \sin^{-1}\left(\frac{E_{dm}}{E_m}\right)$$

$$I_f = \left(\frac{E_f}{R_f}\right)E_{fbase}$$

$$I_a = \sqrt{I_q^2 + I_d^2}$$

$$P_Q = E_{qm} I_d - E_{dm} I_q$$

◊ *Differential equations:*

$$\dot{\delta} = \omega_r - \omega_{grid}$$

$$\dot{E}'_{qm} = [(x'_{dm} - x_{dm})I_d + E_f - E'_{qm}]\tau_{d0}$$

$$\dot{\omega}_r = \frac{[T_m - T_e - (k_d + k_e)\omega_r]}{I}$$

◊ *Outputs :* P_e, P_Q, ω_r

Model 2

◊ *Inputs :* P_{mech}

◊ *Parameters :* $\sigma_p, H_{eq}, D_{eq}, \omega_0$

◊ *States :* ω

◊ *Algebraic equation :*

$P_{el} = f(\overline{P}_{el}, \sigma_p)$ [noise generator]

◊ *Differential equation :*

$$\frac{d\omega}{dt} = \frac{\omega_0}{2H_{eq}}\left[P_{mech} - P_{el} - D_{eq}(\omega - \omega_0)\right]$$

◊ *Outputs :* ω

4.3.7 Gas Merge and Split

4.3.7.1 Introduction

In the Skegton configuration for power plant simulation it is necessary to control the distribution of flows of gas turbine exhaust gases to the waste heat boilers. In practice, this would be achieved by the placing of gas dampers into the flow path. Gas Dampers are similar to valves in that they offer a variable resistance to flow according to the damper position. In this work, individual dampers are not modelled, but the overall effect of splitting the exhaust gas flow for distribution to boilers and the flue is described.

4.3.7.2 Model Assumptions

(i) Friction and other loss neglected.

(ii) Perfect mixing of exhaust gases.

(iii) Dynamics of process and actuators neglected.

(iv) Perfect control is obtained, so that the actual flow to a boiler equals its set point.

4.3.7.3 Model equations

Given that there are two exhaust steams, two boilers and a flue, the following equations apply:

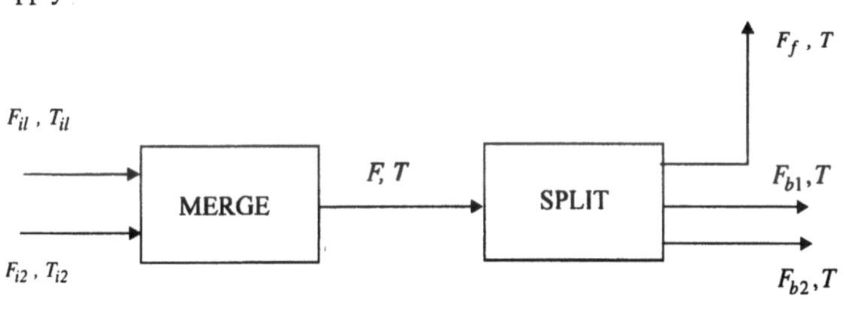

Fig. 4.14 Gas merge and split - block diagram.

Merge

Mass balance : $F = F_{i1} + F_{i2}$

Energy balance : $$h = \frac{F_{i1} h_{i1} + F_{i2} h_{i2}}{F}$$

Split

Mass balance : $F_f + F_{b1} + F_{b2} = F$

where:

F_{i1}, F_{i2} - *incoming exhaust mass flow rates*

h_{i1}, h_{i2} - *incoming exhaust gas enthalpies*

F - *merged mass flow rate*

h - *merged gas enthalpy*

F_f - *flue mass flow rate*

F_{b1}, F_{b2} - *boiler input mass flow rates*

4.3.7.4 Model structure for implementation as an algorithm

◊ *Inputs :* $F_{i1}, F_{i2}, T_{i1}, T_{i2}$ (from gas turbine)

F_{b1set}, F_{b2set} (from supervisory control)

◊ *Algebraic equations:*

$$F = F_{i1} + F_{i2}$$

$$h = \frac{F_{i1} h_{i1} + F_{i2} h_{i2}}{F}$$

$$F_{b1} = F_{b1set}$$

$$F_{b2} = F_{b2set}$$

$$F_f = F - F_{b1} - F_{b2}$$

◊ *Outputs:* F_{b1}, F_{b2}, F_f, h

4.4 Controller Descriptions

4.4.1 Introduction

In order to properly assess the performance of a supervisory plant controller it is important that the lower level component controllers are adequately represented in the power plant simulation. To specify what is in fact an adequate representation requires careful consideration of the interaction that occurs between the supervisory and lower level controls. The supervisor has a direct influence on these low level controllers whose set points it has freedom to select.

4.4.2 Criteria applied to eliminate low level control loops.

The following criteria have been used to eliminate a large number of power plant low level control loops, that are not significant for the development of supervisory control strategies:

- Those low level loops whose action is relatively fast compared to others are neglected. Here it can be assumed that the controlled variable equals the set point at all times. For example control loops associated with valves, actuators, pumps etc.
- Those low level loops that are included for purely safety considerations are neglected. For example duplicated, redundant speed/load control loops on turbines etc. It was not within the scope of the current project work to model safety control systems.
- Where there exists a supervisory control action in conventional power plant control system architectures, this has been neglected. This will be replaced by the total plant supervisory control action to be defined in future research work.

  For example in the control of boiler-turbine units, supervisory control modes known as *boiler following* or *turbine following* are typically implemented in industrial power plants.
- Those low level loops whose set point is fixed, thus presenting no extra degrees of freedom at the supervisory level, are neglected. It should be noted that in such cases, the component models are constructed to capture the closed loop dynamics response. Here the effects of the low level control action on the component dynamic response are described by the model without representing the low level control structure as a separate entity.

In what follows the low level control loops for each component model that are not excluded by reason of the above criteria, are described.

4.4.3 Boiler Control Configuration

4.4.3.1 Introduction

The boiler control system in real power plant applications consist of hundreds of local control loops maintaining values of different variables at different parts of the boiler. However, for the purpose of this simulation, only the most significant loops are to be represented. Consequently, the boiler model described in Section 4.3.1 includes only four manipulated inputs - which are adjusted locally by the boiler control system. The remaining control loops are assumed to be fast enough and accurate enough (which is in practice the case), not to need to be modelled. The control system chosen is the most common approach in which four independent single input single output controllers are applied. The assignment of input and output variables for the boiler control are as follows:

- superheated steam pressure is regulated by adjusting the fuel flow
- drum water level is regulated by adjusting the feedwater flow
- superheated steam temperature is regulated by the attemporation flow
- furnace air pressure is regulated by the air flow to the furnace.

4.4.3.2 Literature Review

Configurations of the boiler control system may be found in the following papers : [*Waddington and Maples, 1987*], [*Amin et al, 1984*], [*Keeton, 1984*], [*Yang and Hogg, 1992*], [*Achluvalia and Domenichini, 1990*], [*Dwarakanath et al, 1982*], [*Welfonder, 1992*], [*Mann and Lauster, 1992*], [*IEEE, 1992*].

The configuration proposed is mainly based on the paper [*Welfonder, 1992*] with some justification for different control modes from the paper of [*IEEE, 1991*].

4.4.3.3 Block diagram

The block diagram of the boiler control system is presented in Fig. 4.15. The diagram consists of four feedback proportional integral controllers (PI) and one feedforward proportional derivative controller (PD). The nonlinear block converts the value of the desired output power P_{DES} to the value of superheated steam pressure set point.

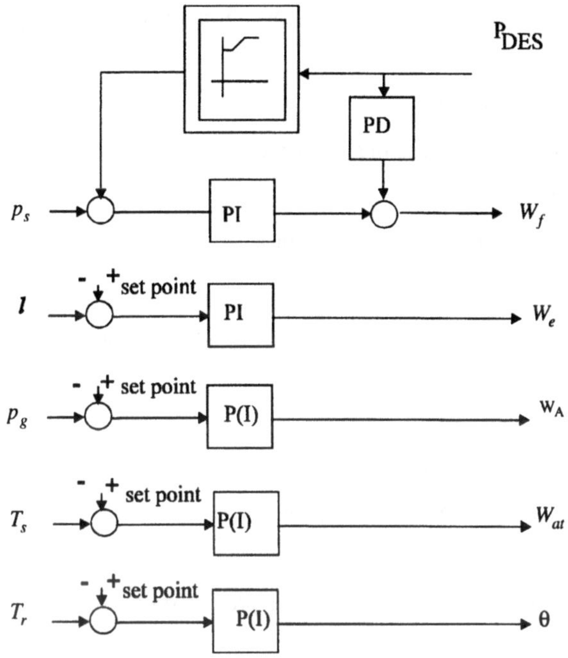

Fig. 4.15 Boiler control system - block diagram. Key : input signals to the controller : p_s - *superheated steam pressure,* l - *drum water level,* p_g - *furnace air pressure,* T_s - *superheated steam temperature,* T_r - *reheated steam temperature;* output signals from the controller : w_F - *fuel flow to the furnace,* w_e - *feedwater flow,* w_A - *air flow to the furnace,* w_{at} - *attemporation water flow,* Θ - *tilt angle.*

4.4.4 Gas turbine control configuration

4.4.4.1 Introduction

The gas turbine controller regulates both the gas turbine and the gas turbine generator. One of the most popular design is at present the system *SPEEDTRONIC* developed for General Electric gas turbines. For the purpose of this project only modulating control of mechanical side of the gas turbine is of interest. Thus the model, based on *SPEEDTRONIC* Mark 4, is significantly simplified.

The simplified model of the gas turbine controller consists of four inputs and three outputs. Inputs to the controller, which are outputs from the gas turbine model, are:

- delivered mechanical power
- speed of rotation (related to electrical frequency)
- outlet gas temperature
- outlet gas composition (NO_X and CO content).

Outputs from the controller which are inputs to the gas turbine model are:

- fuel flow to the combustor
- air flow (inlet guide vanes position)
- steam (or water) injection flow to the combustor

The assignment of input and output variables is presented on the block diagram in Section 4.4.4.3.

4.4.4.2 Literature Review

The concept of the gas turbine control system which is applied in this report is based on the *SPEEDTRONIC* Mark 4 description as presented in the paper *Rowen* [1988]. Some considerations concerning the subject may be also found in *Yacobucci*, [1991], and *Polimeros*, [1981].

3.3.3. ***Block diagram***

The block diagram of the gas turbine control system is presented in Fig. 4.16. The diagram consists of four PID feedback controllers. LVG stands for *Least Value Gate* which transmits the minimum of three incoming signals. MVG stand for *Maximum Value Gate* and transmits the maximum of two signal, one of which is minimum allowable fuel flow.

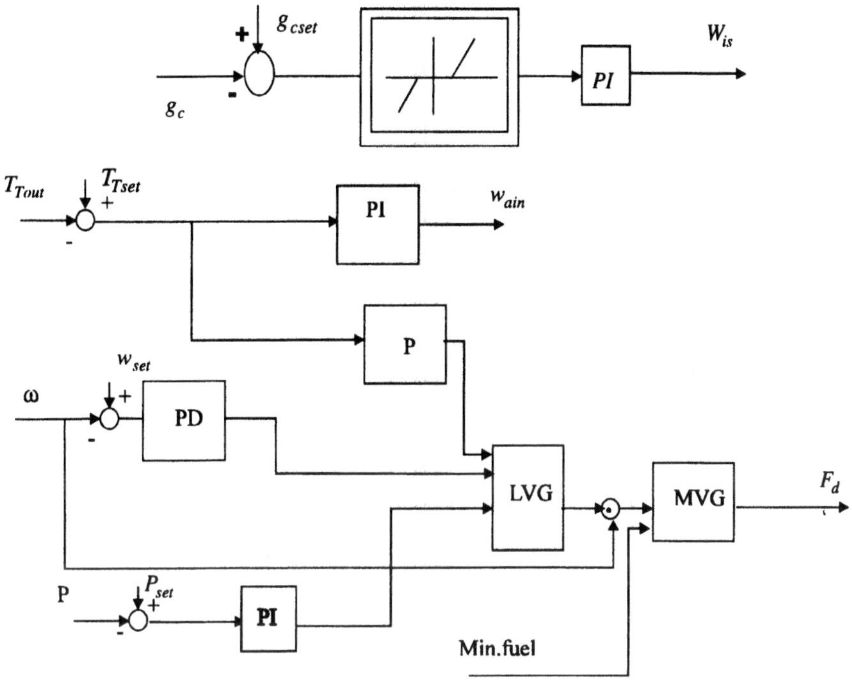

Fig. 4.16 Gas turbine control system - block diagram.

The input signals to the controller are:

- P - *power delivered by the turbine*
- ω - *electrical frequency of the generator*
- T_{Tout} - *outlet gas temperature*
- g_{CNOx} and g_{CCO} - *NO_x and CO content in the outlet gas*

Outputs from the controller are:

- F_d - *fuel flow desired value*
- w_{ain} - *air flow (inlet guide vanes position)*
- w_{is} - *steam (or water) injection flow.*

4.4.5 Steam turbine control configuration

4.4.5.1 Introduction

The steam turbine controller regulates the steam turbine, the steam turbine generator and the steam turbine admission valves. The design reported enables different modes of control to be imposed on the steam part of the plant, namely:

- turbine following mode
- boiler following mode
- unit coordinated mode.

This may be realised by appropriate adjustments of set points by a supervisory controller.

The inputs to the steam turbine controller, which are outputs from the steam turbine the generator and the admission valves models are:

- delivered mechanical power
- electrical frequency (rotation speed of the generator)
- steam pressure at the inlet to the turbine.

Outputs from the controller are:

- throttle valve position
- bypass valve position

The connections between input and output variables are presented on the block diagram in the Section 4.4.5.3.

4.4.5.2 Literature review

The control system presented in this report is based on the paper [*IEEE Committee, 1991*] with some modifications from the paper [*Welfonder, 1992*]. Some the remarks may be found in the papers : [*Ahluvalia and Domemchini, 1990*], [*Kure-Jensen and Hanisch, 1991*].

4.4.5.3 Block diagram

The block diagram of the steam turbine control system is presented in Fig. 4.17. The diagram consists of two controllers : proportional frequency controller and integral power controller. The nonlinear blocks realise the contrained control concept by opening or closing valves when the steam pressure approaches allowable limits. The inputs to the controller are :

- P - *delivered mechanical power*
- ω - *electrical frequency*
- p_{sbis} - *steam pressure at the inlet to the turbine (outlet from the throttle valve)*

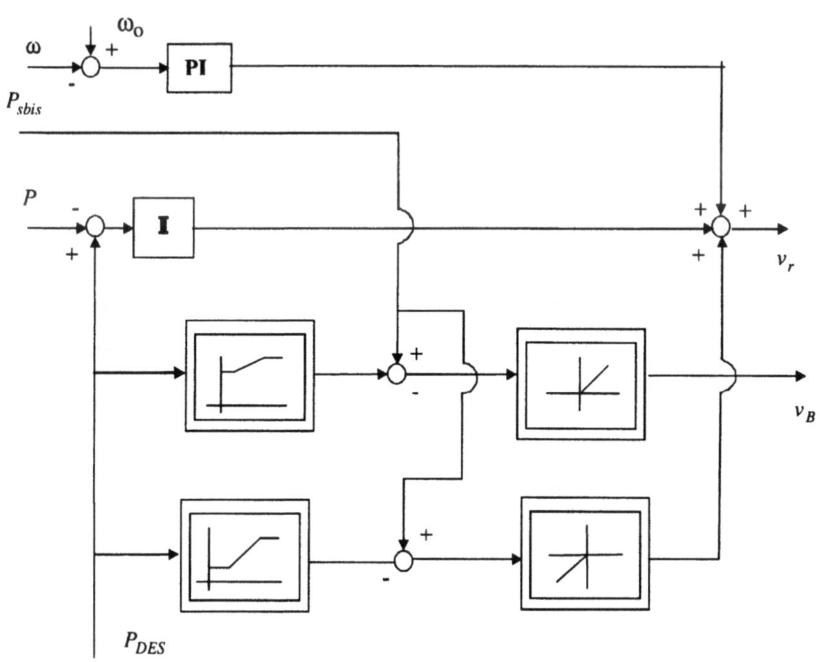

Fig. 4.17 Steam Turbine Control System - block diagram.

Outputs signals from the controller are denoted as follows:

- υ_T - *throttle valve position*
- υ_B - *bypass valve position*

4.4.6 Condenser control configuration

4.4.6.1 Introduction

The condenser is considered to have only one manipulated (control) variable which is the cooling liquid flow. The controlled variable is condensate exit temperature. The control system is very simple consisting of only one P(ID) controller.

4.4.6.2 Literature review

Some basic control configurations for condensers are discussed by [*Roffel and Rijnsdorp, 1982*].

4.4.6.3 Block diagram

The block diagram of the condenser controller is presented in Fig. 4.18. T_{in} denotes the input signal to the controller which is condensate temperature. w_f denotes cooling liquid mass flow.

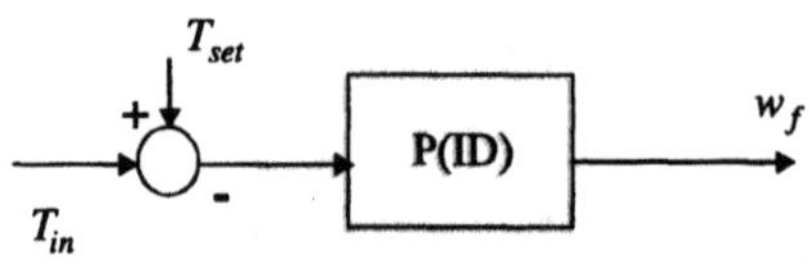

Fig. 4.18 Condenser controller - block diagram.

4.4.7 Feedwater system control configuration

4.4.7.1 Introduction

Two primary control objectives exist for the feedwater system; to maintain the vessel pressure and liquid level in the deaerator at design values. At saturation conditions the steam pressure inside the deaerator determines the outlet feedwater temperature and the amount of residue oxygen within it. Steam pressure is regulated by adjusting the flow of extraction steam supplied to the deaerator. Similarly liquid level is regulated via manipulation of inlet makeup water flow.

4.4.7.2 Block Diagram

The block diagram of the feedwater system controller is presented in Fig. 4.19.

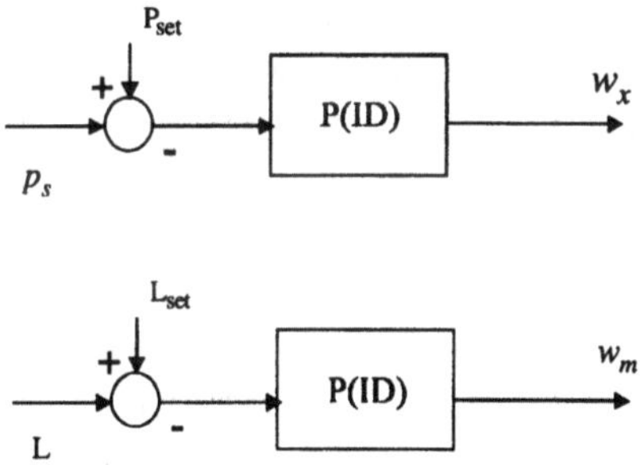

Fig. 4.19 Feedwater system controller - block diagram.

4.5 Conclusions

In this chapter mathematical models have been presented of each CC/CHP power plant component, as contained in the Skegton generic simulation configuration.

In each case a literature review has been followed by a list of modelling assumptions, before presenting the model equations. For the more complex components a block diagram depicting the interactions between sub-modules has been included. Thc model equations are then reformulated as an algorithm for implementation. For the major plant components, conventional control schemes have been described.

4.6 References for the Chapter

Ahluvalia K.S. and R. Domenichini ***Dynamic modelling of a combined cycle plant.*** Trans. ASME Journal of Engineering for Gas Turbines and Power, Vol. 112, 164-167, Apr. 1990

Amin M., G.Y. Masada and B.F. Womack ***Application of a multivariable self-tuning controller to a power plant boiler.*** Proc. American Control Conference, 1984

Arrillaga J. and C.P. Arnold ***Computer Analysis of Power Systems.*** John Wiley & Sons, 1990

Åström K.J. and R.D.Bell ***Simple drum boiler models.*** Proc. IFAC Power Systems Modelling and Control Applications, Brussels, 1988

Balchen J.G. and K.I. Mummé ***Process Control Structures and Applications.*** Van Nostrand Reinhold Company Inc., 1988

Biss D. and M.J. Grimble ***Robust control of a 1.5MW free turbine with complex load: linear controller designs.*** Submitted to Control Theory Applications, Engineering Systems Design and Analysis, ASME-94, 1993

Cegrell T. ***Power System Control Technology.*** Prentice Hall International, 1986

Chien K.L., E.I. Ergin, C. Ling and Allyn Lee ***Dynamic analysis of a boiler.*** Transactions of the ASME, November 1958

Cheres E. ***Small and medium size drum boiler models suitable for long term dynamic response.*** IEEE Transactions on Energy Conversion, Vol.5, No.4, December 1990

Chiacciarini H.G., A.C. Desages, J.A. Romagnoli and A. Palazoglu ***Variable structure control strategies : application to a MIMO nonlinear steam generating unit.*** Proc. American Control Conference, 1992

Cohen H., G.F.C. Rogers and H.I.H. Saravanamuttoo **Gas Turbine Theory (3rd edition).** Longman, 1987

Cori R. and T. Busi ***Parameter identification of a drum boiler power plant.*** Proc. 3rd Power Plants Dynamics, Control and Testing Symposium, Knoxville, Tennessee, September 7-9, 1977

Daniel K.J., A.W. Crapo and, D.H. Brown ***Transient simulation and experimental verification of gas-steam and liquid-steam, shell, and tube type heat exchangers.*** Proc. American Control Conference, 1984

Dwarakanath M.H., B. Dembart, A.M. Erisman, K. Hemmaplardh and J.W. Manke ***A generalized methodology for modelling system components in power system***

dynamic simulation. IEEE Trans. on Power Apparatus and Systems, Vol. PAS-101, No.1, 136-146, Jan.1982

Fawke A.J., H.I.H. Savanamuttoo and M Holmes ***Experimental verification of a digital computer simulation method for predicting gas turbine dynamic behaviour.*** Proceedings the Institution of Mechanical Engineers, Vol.186, No.27, 1972

Ferrarini L. and C. Maffezzoni ***Simulation of heat exchangers in fossil fired power plants : a method to avoid Gibbs' phenomena.*** Proceedings of the IFAC Symposium on Control of Power Plants and Power Systems, Munich, 1992

Gunn, D. and Horton, R., ***Industrial Boilers,*** Longman, 1989.

Hussain A. and H. Seifi ***Dynamic modelling of a single shaft gas turbine.*** Proceedings of the IFAC Symposium on Control of Power Plants and Power Systems, Munich, 1992

Hung W.W. ***Dynamic simulation of gas-turbine generating unit.*** IEE Proceedings-C, Vol.138, No.4, July 1991

IEEE Committee ***Dynamic models for fossil fueled steam units in power system studies.*** IEEE Trans. on Power Systems, Vol.6, No.2, 753-761, May 1991

IEEE Committee ***Dynamic models for steam and hydro turbines in power system studies.*** IEEE Committee Report, IEE PES Winter Meeting, New York, 1972

Kapadia M. and C.H. Wolgemuth ***A dynamic model of a condenser in a closed Rankine cycle power plant.*** Proc. American Control Conference, 1984

Keeton J.M. ***Advanced control techniques applied to a fossil power plant.*** Proc. American Control Conference, 1984

Knowles J.B. ***Simulation and control of electrical power stations.*** John Wiley & Sons, 1990

Kure-Jensen J. and R. Hanisch ***Integration of steam turbine controls into power plant systems.*** IEEE Trans. on Energy Conversion, Vol.6, No.1, 177-185, March 1991

McDonald J.P. and H.G. Kwatny ***A mathematical model for reheat boiler-turbine-generator systems.*** IEEE Paper 70 CP221-PWR, PES Winter Power Meeting, New York, January 25-30, 1970

Maffezzoni C. ***Issues in modelling and simulation of power plants.*** Proceedings of the IFAC Symposium on Control of Power Plants and Power Systems, Munich, 1992

Mann J. and G.K. Lausterer ***Temperature control using state feedback in a fossil fired power plant.*** Proceedings of the IFAC Symposium on Control of Power Plants and Power Systems, Munich, 1992

de Mello F.P. ***Boiler models for system dynamic performance studies.*** IEEE Trans. on Power Systems, Vol.6, No.1, 66-74, Feb. 1991

de Mello F.P., R.J. Mills and W.F. B'Rells ***Automatic generation control, part I: Process modeling, part II: digital control techniques.*** IEE Transactions on Power Apparatus and Systems, Vol.PAS-92, No.2, pp. 710-725, March/April 1973

Nicholson H. ***Dynamic optimisation of a boiler.*** Proc. IEE Vol.111, No.8, August 1964

Onions R.A. and A.M. Foss ***Improvements in the dynamic simulation of gas turbines.*** source unknown, around 1986

Polimeros G. ***Energy Cogeneration Handbook.*** Industrial Press Inc., 1981

Rafian M., M.J.H. Sterling and M.R. Irving ***Real-time power system simulation.*** IEE Proceedings Pt.-C, Vol.134, No.3, pp. 206-223, May 1987

Ray A. ***Dynamic modelling of power plant turbines for controller design.*** Applied Mathematical Modelling, Vol.4, April 1980

Ray K.S. and Majumder D.D. ***Fuzzy logic control of a nonlinear multivariable steam generating unit using decoupling theory.*** IEEE Transactions on Systems, Man, and Cybernetics, Vol. SMC-15, No.4, July/August 1985

Rhine J.M. and R.J. Tucker ***Modelling of gas-fired furnaces and boilers.*** McGraw-Hill, London,1991

Roffel B. and Rijnsdorp J.E. ***Process Dynamics, Control and Protection.*** Ann Arbor Science Publishers, Ann Arbor, Michigan, 1982

Rogers G.F.C. and Y.R. Mayhew ***Engineering Thermodynamics.*** Longman, 1992

Rowen W.I. ***Simplified mathematical representations of heavy-duty gas turbines.*** ASME Journal of Engineering for Power, 865-871, Oct. 1983

Rowen W.I. ***Speedtronic Mark IV control system.*** Alsthom Gas Turbine Reference Library (AGTR 880), 1988

Shinskey F.G. ***Process Control Systems.*** McGraw-Hill, 1979

Shobeiri T. ***Digital computer simulation of the dynamic operating behaviour of gas turbines.*** Brown Boveri Review, 3-87, 1987

Sorensen H.A. ***Energy Conversion Systems.*** John Wiley & Sons, 1983

Stephanopoulos G. ***Chemical Process Control, an Introduction to Theory and Practice.*** Prentice Hall, 1984

Thermoflow Inc. (ed.) ***Design of gas turbine combined cycle and cogeneration systems.*** Lecturenotes from the course in St Albans, England, July 31-August 2, 1991

Thomas Y., Ph. de Larminat, D. Sarlat, A. Barraud, J. Descusse and J.M. Piasco ***Analysis of a DDC experiment based on state space methods.*** Transactions of ASME, September 1976

Waddington J. and G.C. Maples ***The control of large coal- and oil- fired generating units.*** Power Engineering Journal, 25-34, Jan.1987

Welfonder E. ***Constrained control concepts in power plants and power systems for avoiding emergency conditions.*** Proceedings of the IFAC Symposium on Control of Power Plants and Power Systems, Munich, 1992

Wood A.J. and B.F. Wollenberg ***Power Generation, Operation and Control.*** John Wiley & Sons, 1984

Yacobucci R.B. ***A control system retrofit for a GE frame 5 turbine/generator unit.*** IEEE Trans. on Energy Conversion, Vol.6, No.2, 225-230, June 1991

Yang P. and B.W. Hogg ***Continuous-time generalized predictive control of a boiler model.*** Proceedings of the IFAC Symposium on Control of Power Plants and Power Systems, Munich, 1992

-- ***Bibliography of literature on steam turbine-generator control systems***, IEEE Trans. on Power Apparatus and Systems, Vol. PAS-102, pp.2959-2970, Sept. 1983

Update of Bibliography of literature on steam turbine-generator control systems, IEEE Committee Report, Trans. on Energy Conversion, Vol.3, No.3, pp. 560-567, Sept. 1988

CHAPTER 5

SIMULATION STUDIES

5.1 Introduction.

This chapter describes the library of modules for Combined Cycle and Combined Heat and Power plant simulation as implemented in $MATRIX_X$. The modules may be used stand-alone or may be configured to a given power plant structure. Then, some simulation experiments performed for the case study Skegton unit configuration are presented.

The modules in general are formulated in two different ways. The first is as FORTRAN subroutines, the second is as standard blocks from the $MATRIX_X$ library. In a few cases modules can be represented as a composite of the above.

To give an idea of the size of the Skegton simulation; it has in total, 132 dynamic states and more than 400 algebraic equations.

5.2 Module Library Description

5.2.1 The Boiler

The boiler is constructed from the integration of the following modules:

- Five FORTRAN Subroutines :
  - furnace.f
  - drum.f
  - risers.f
  - superheater.f
  - reheater.f
- superblock designed using standard $MATRIX_x$ library blocks
  - boiler - control

The overall structure of the boiler model is depicted in Fig. 5.1. Fig. 5.2 gives an insight into the internal structure of the superblock BOILER. A short description of each of the above modules now follows.

5.2.1.1 The Furnace.

The module comprises a MATRIXx model of a boiler furnace. The model consists of 2 differential equations and 13 algebraic equations. The combustion temperature is calculated from gas tables using linear approximation.

5.2.1.2 The Drum Part.

The module consist of two superblocks : DRUM and RISERS. The drum model includes three differential equations and 16 algebraic equations. Saturated steam tables and saturated water tables are used to calculate the steam and water conditions. A polynomial fit is applied to

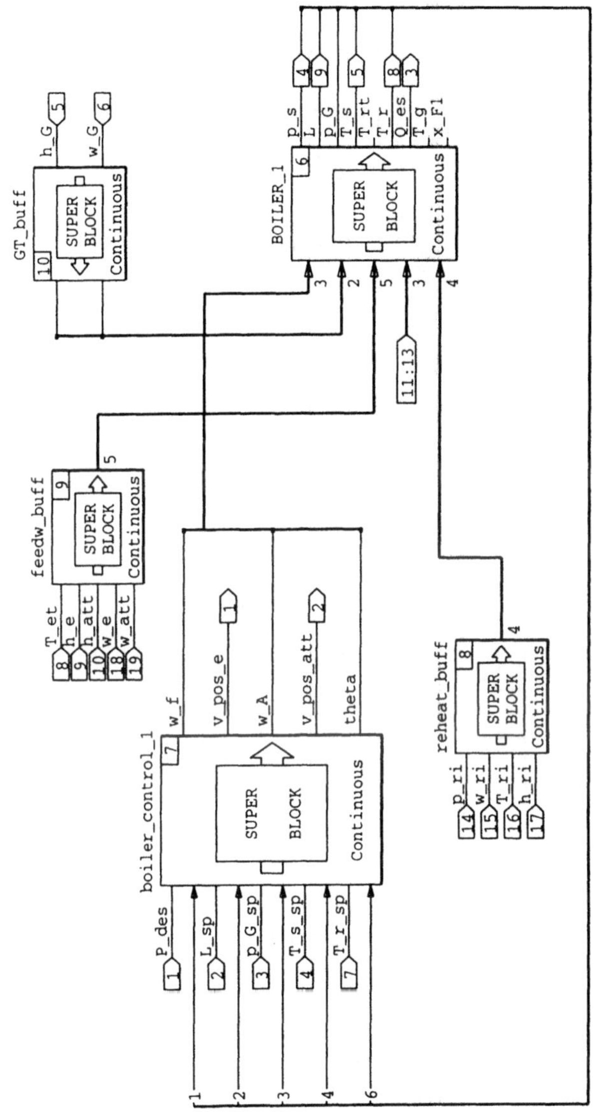

Fig. 5.1 Boiler system model structure.

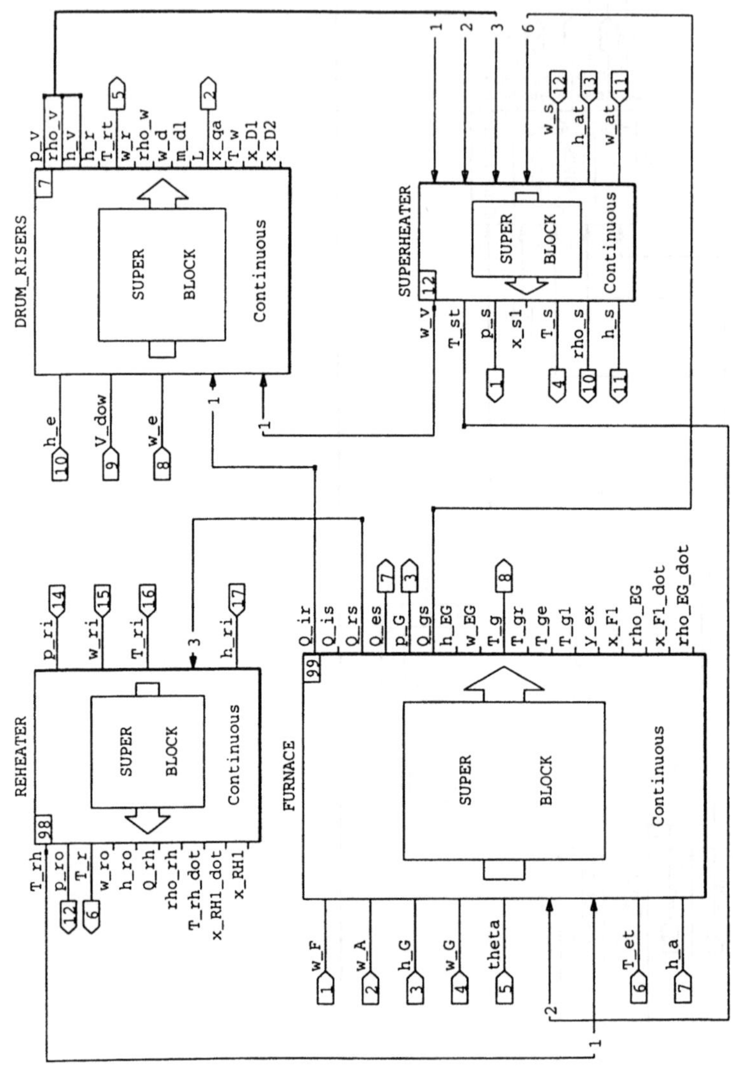

Fig. 5.2 Superblock BOILER - internal structure.

approximate steam tables relations. The code gives a choice between a spherical or cylindric shape for the drum. The riser model consists of three differential equations and three algebraic equations.

5.2.1.3. Superheater plus Attemporator.

The superheater model consist of three differential equations and six algebraic equations. The superheated steam conditions are calculated using the ideal gas assumption.

5.2.1.4. The Reheater.

The model includes three differential equations and six algebraic equations. Relations for superheated steam are approximated using the ideal gas assumption.

5.2.1.5. Boiler Control.

This module includes standard MATRIXx blocks. It provides low level control of boiler superheated steam pressure, drum water level, furnace air pressure, superheated steam temperature and reheated steam temperature. The block diagram presented in Fig. 5.3 shows the MATRIXx implementation of the model.

5.2.2 The Gas Turbine

The gas turbine is constructed from the integration of the following modules:

- Three FORTRAN subroutines

    GT_compress.f

    GT_combust.f

    GT_turb.f

- Three superblocks defined using standard MATRIX_X library blocks

    GT_control

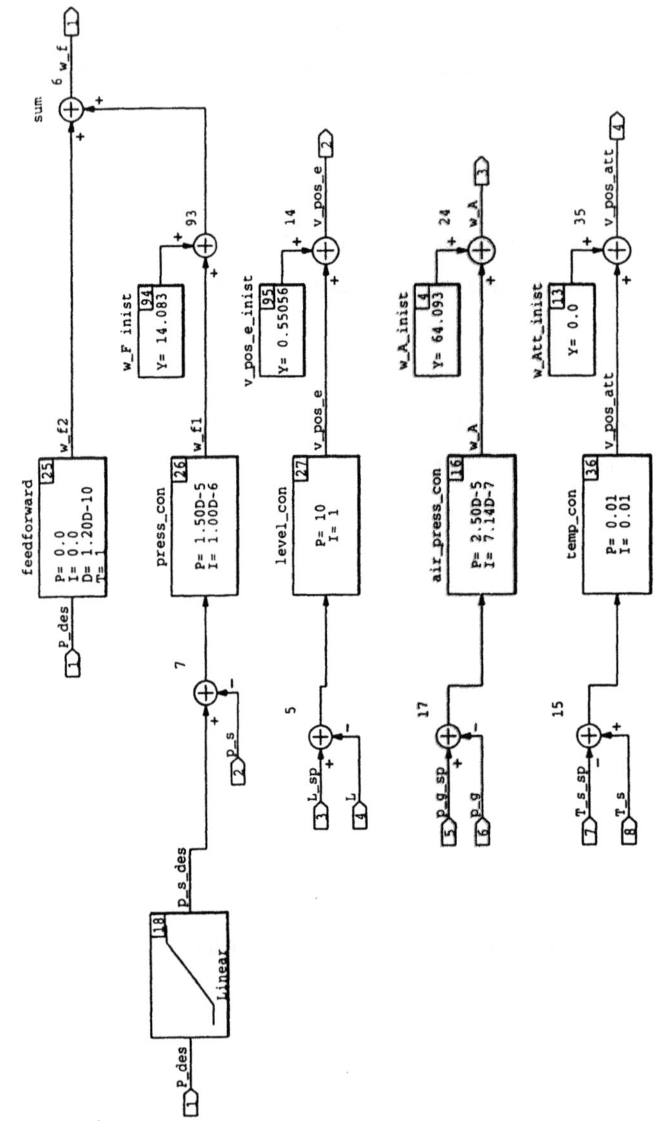

Figure 5.3. The Boiler Controller

GT_fuelsys

turbogen_GT.

The overall structure of the gas turbine model is depicted in Fig. 5.4. Fig. 5.5 gives insight into the internal structure of the superblock GAS_TURBINE. The description of each of the above modules now follows.

5.2.2.1 Fuel System

This module implements the model for the fuel valve and associated fuel system dynamics.

5.2.2.2 Compressor

The gas-turbine compressor is modelled using standard one dimensional steady gas flow equations. The models consists of 10 algebraic equations.

5.2.2.3 Combustion Chamber

The gas-turbine combustor is modelled using steady state energy balance for combustion at reference temperature. The models consists of eight algebraic equations.

5.2.2.4 The Turbine

The gas-turbine turbine section is modelled using standard one dimensional steady gas flow equations. The model consists of 13 algebraic equations.

5.2.2.5 The GasTurbine Controller

This module is a low level controller for regulation of GT power, speed, exhaust gas, temperature and pollution. Fig. 5.6 shows the MATRIXx implementation of the model.

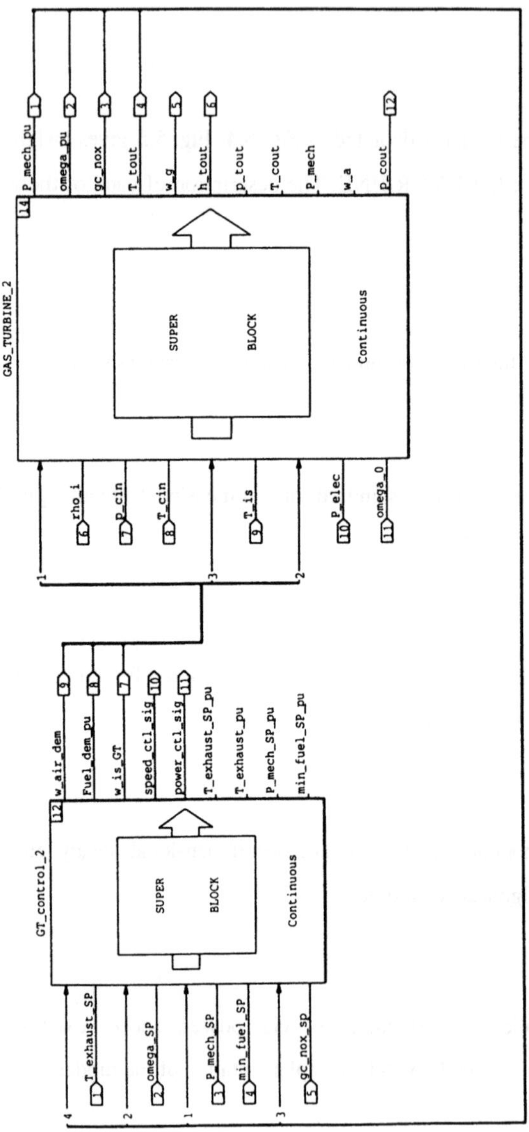

Fig. 5.4 Gas turbine system.

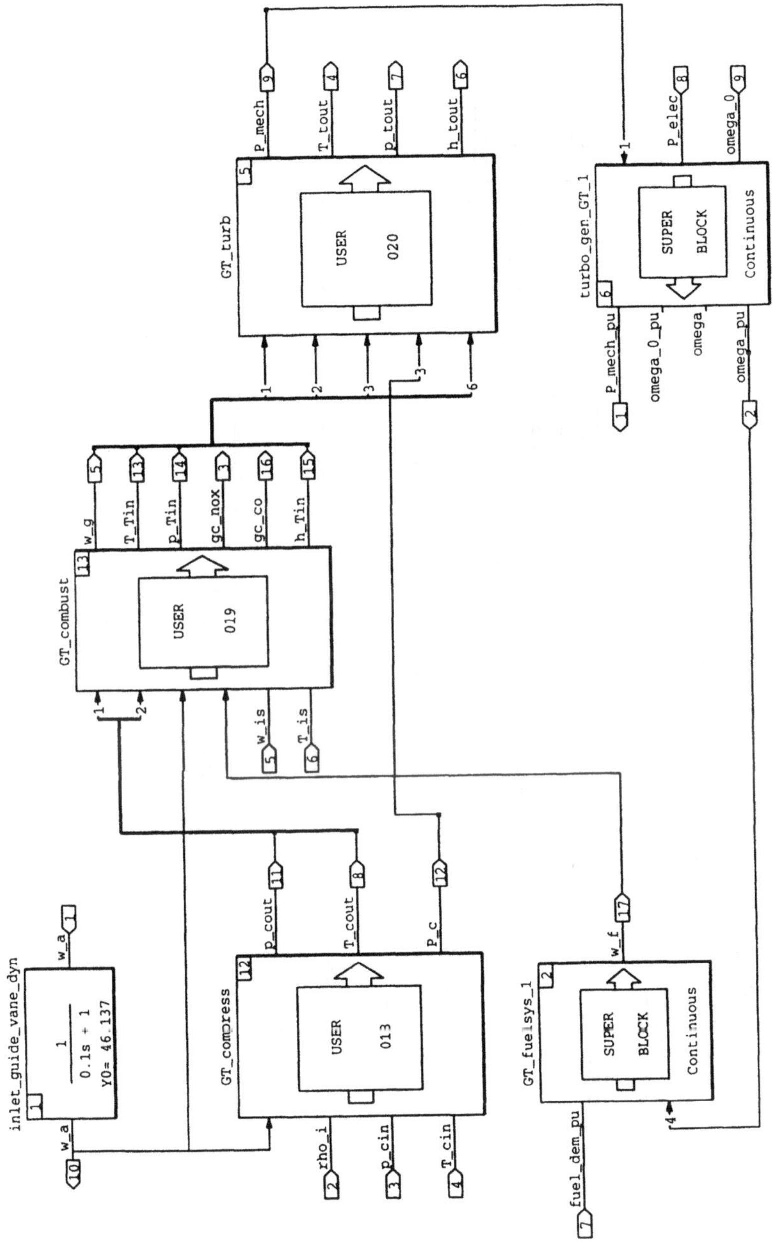

Fig. 5.5 Superblock GAS_TURBINE - internal structure.

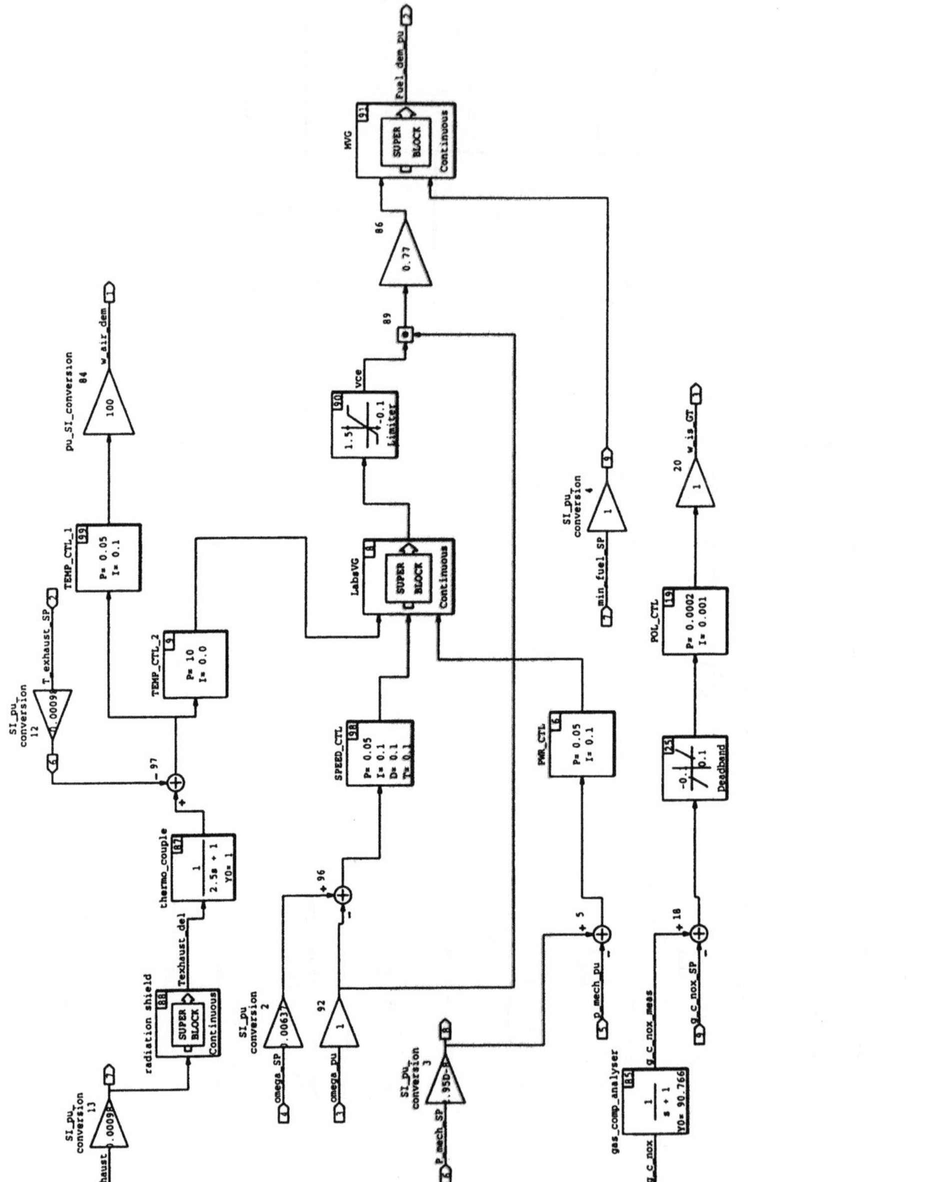

Fig. 5.6 Gas turbine controller.

5.2.3 The Steam Turbine

The steam turbine is constructed from the integration of the following modules:

- Four FORTRAN subroutines

  ST_hpsec.f

  ST_ipsec.f

  ST_lpsec.f

  Valve-gas.f

- Three superblocks defined using standard $MATRIX_X$ library block

  ST_control

  SGS (speed governor system)

  Turbo_gen_GT.

The overall structure of the steam turbine module is depicted in Fig. 5.7. Fig. 5.8 gives an insight into the internal structure of the superblock STURB. The short description of each of the above modules now follows.

5.2.3.1 The High Pressure Section

The steam-turbine HP turbine section is modelled using standard one dimensional steady gas flow equations. The model consists of 15 algebraic equations and three first order differential equations.

5.2.3.2 The Intermediate Pressure Section

The steam turbine IP turbine is modelled using standard one dimensional steady gas flow equations. Note that there is no modelled steam storage since this is already represented in the boiler model as the reheater. The model consists of 13 algebraic equations.

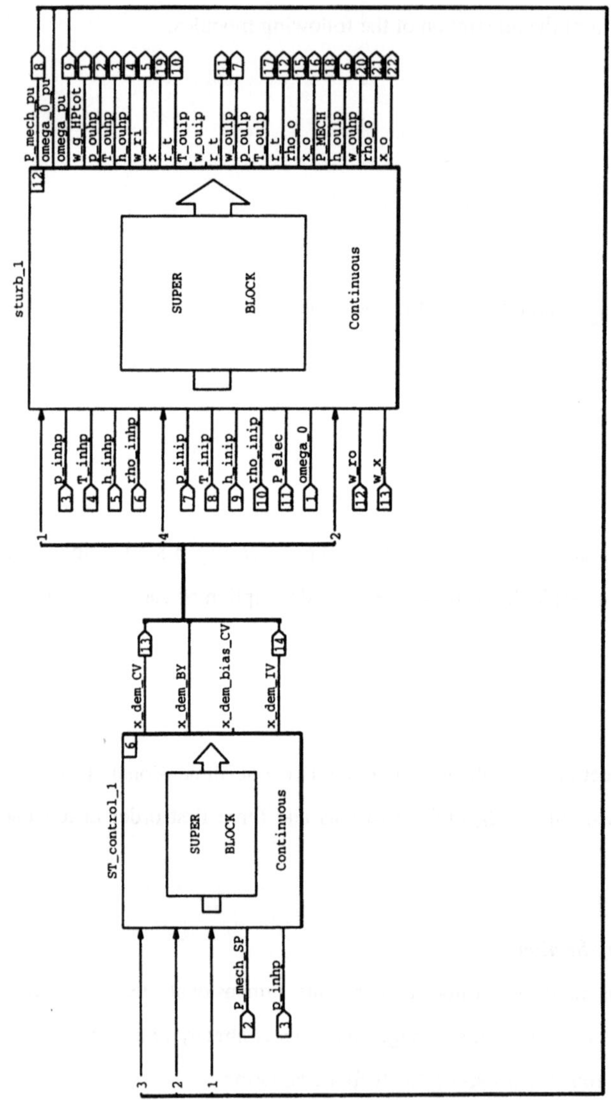

Fig. 5.7 Steam turbine model.

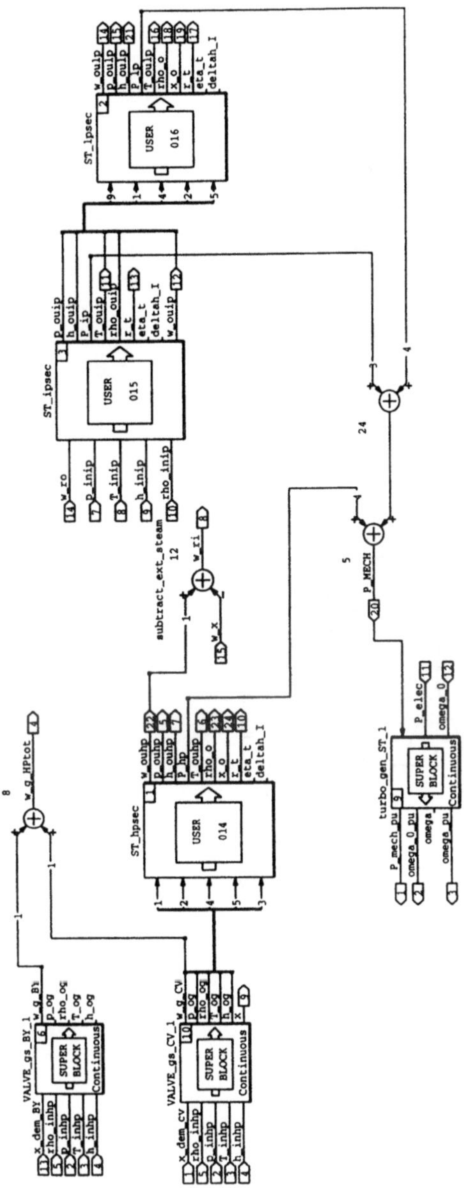

Fig. 5.8 Superblock STURB - internal structure.

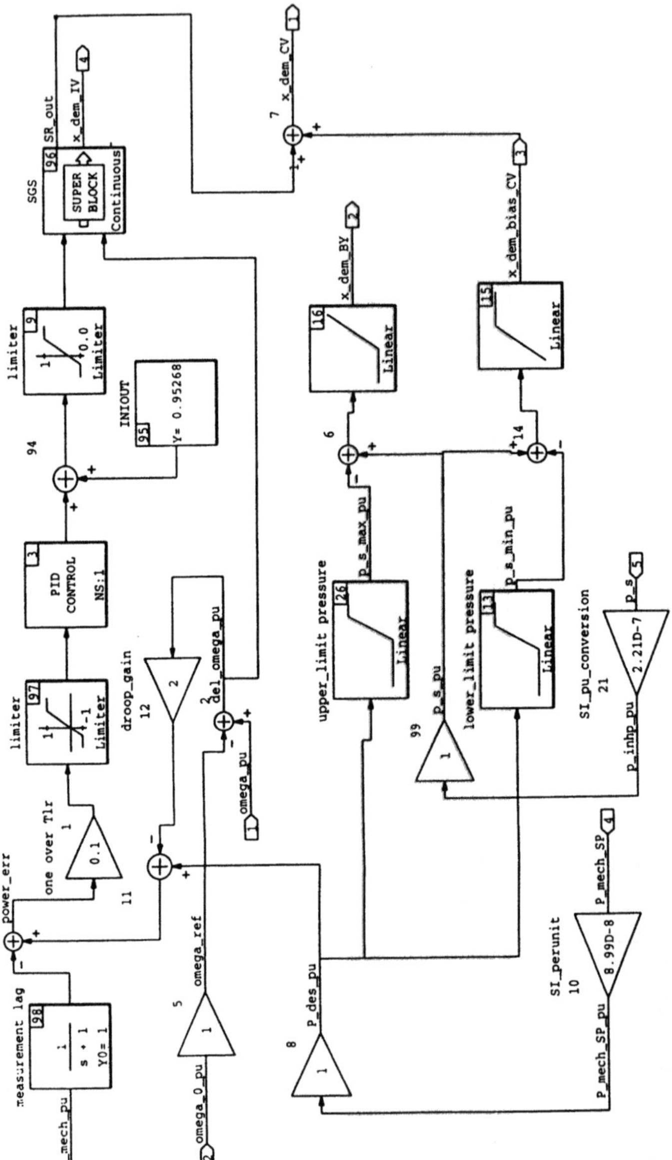

Fig. 5.9 Steam turbine controller.

5.2.3.3 The Low Pressure Section

The steam-turbine LP turbine section is modelled using standard one dimensional steady gas flow equations The model consists of 15 algebraic equations and three first order differential equations.

5.2.3.4 The Steam Turbine Controller

This module provides low level control of steam-turbine power and speed. Fig. 5.9 shows the $MATRIX_X$ implementation of the model.

5.2.4 The Condenser

The condenser is constructed from the integration of the four FORTRAN subroutines :

- Shell.f
- Steam.f
- Tube.f
- Liquid.f

The condenser controller is a simple single loop temperature control and is implemented using standard $MATRIX_X$ library blocks. The overall structure of the condenser model is depicted in Fig. 5.10. The superblock CONDENSER is depicted in Fig. 5.11. The description of each of the above modules now follows.

5.2.4.1 Shell Material Equations

The module describes shell metal equations for the condenser. The model consists of one differential equation and one algebraic equation.

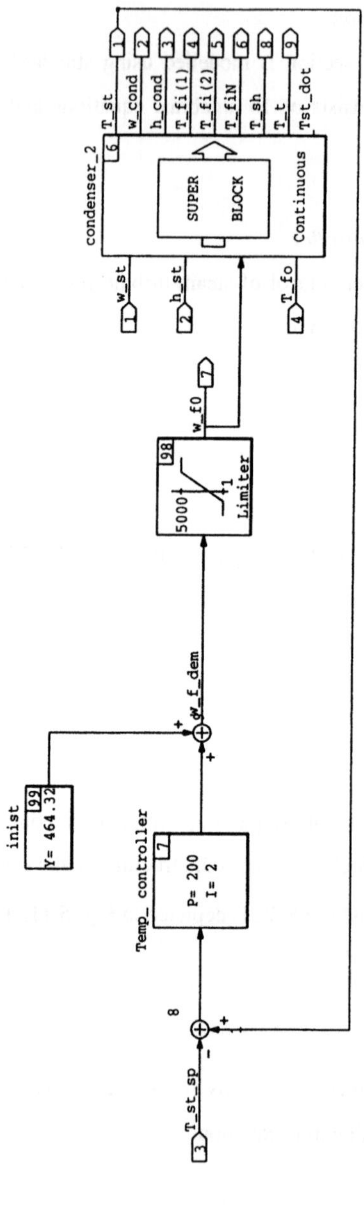

Fig. 5.10 Condenser system - model structure.

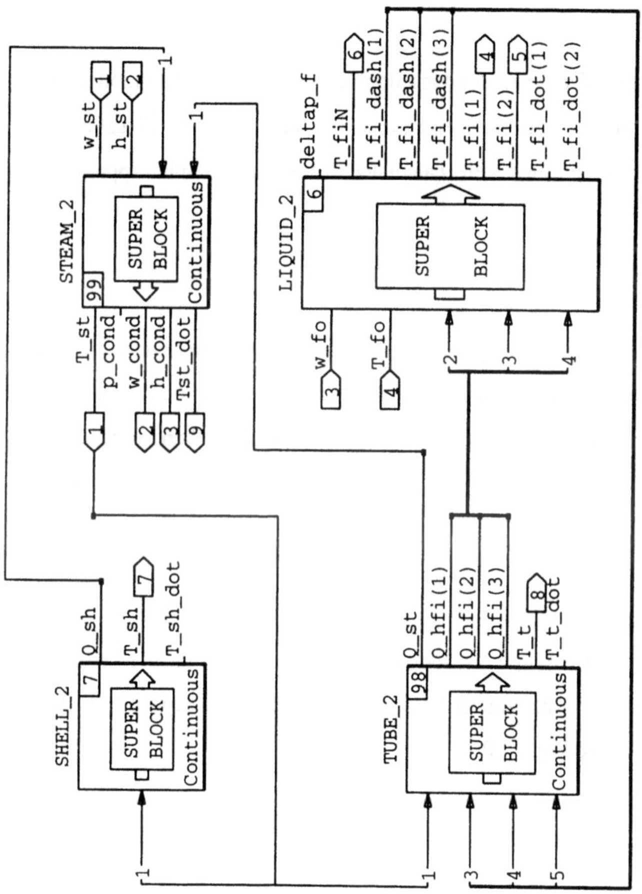

Fig. 5.11 Superblock CONDENSER - internal structure.

5.2.4.2 Tube Material Equations

The module represents heat exchange for metal tubes inside the condenser. The model consists of one differential equation and 4+N algebraic equations (where N is number of sections in the cooling liquid model). In this implementation N was set to three.

5.2.4.3 Steam Equations

The module provides equations for steam condensation in the condenser. The model consists of one differential equation and nine algebraic equations. Partial derivatives of saturated steam parameters with regard to temperature were calculated from a polynomial approximation of steam tables.

5.2.4.4 Liquid Equations

This module represents equations of cooling liquid as it flows through the condenser. The liquid model is divided into N sections. (In this implementation N=3). Each section is described by one differential equation and two algebraic equations. One more equation ; thermal capacitance c_fi is assumed to be the same for each section. Finally, the overall pressure drop for the liquid is calculated.

5.2.4.5 Condenser Controller

The condenser controller consists of only one loop regulating the temperature of condensate by adjusting the cooling water flow. The standard PI controller is used as depicted on the overall diagram of condenser system (see Fig. 5.10).

5.2.5 The Feedwater System

The structure of the feedwater system is presented in Fig. 5.12. Fig.5.13 gives an insight into the internal structure of the superblock FEEDWATER. The feedwater controller consists of two PID loops. The feedwater model is constructed from six superblocks :

- Deaerator
- Pump
- Economiser
- Two Valve Blocks

The description of each of the above modules now follows.

5.2.5.1 The Economizer

The module includes three differential equations and five algebraic equations. The model is similar to the reheater model of the boiler.

5.2.5.2 The Deaerator

The $MATRIX_X$ model of a deaerator - feedwater tank consists of two differential equations and seven algebraic equations. The water in the deaerator is assumed to be in saturated conditions. A polynomial fit is used to approximate saturated water tables.

5.2.5.3 The Pump

The module uses a simple model of a pump. The model contains of seven algebraic equations. A first order lag model the actuation.

5.2.5.4 The Valve (liquid)

This is a simple model of a valve for liquid flows. The model contains of two algebraic equations.

5.2.5.5 Feedwater System Controller

This module provides low level control to deaerator water level and deaerator water pressure. The block diagram presented in Fig. 5.14 shows $MATRIX_X$ implementation of the model.

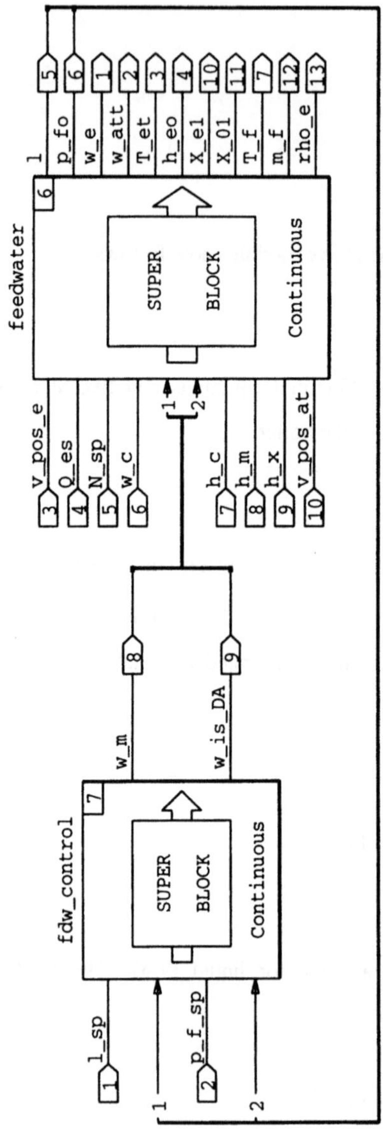

Fig. 5.12 Feedwater system - model structure.

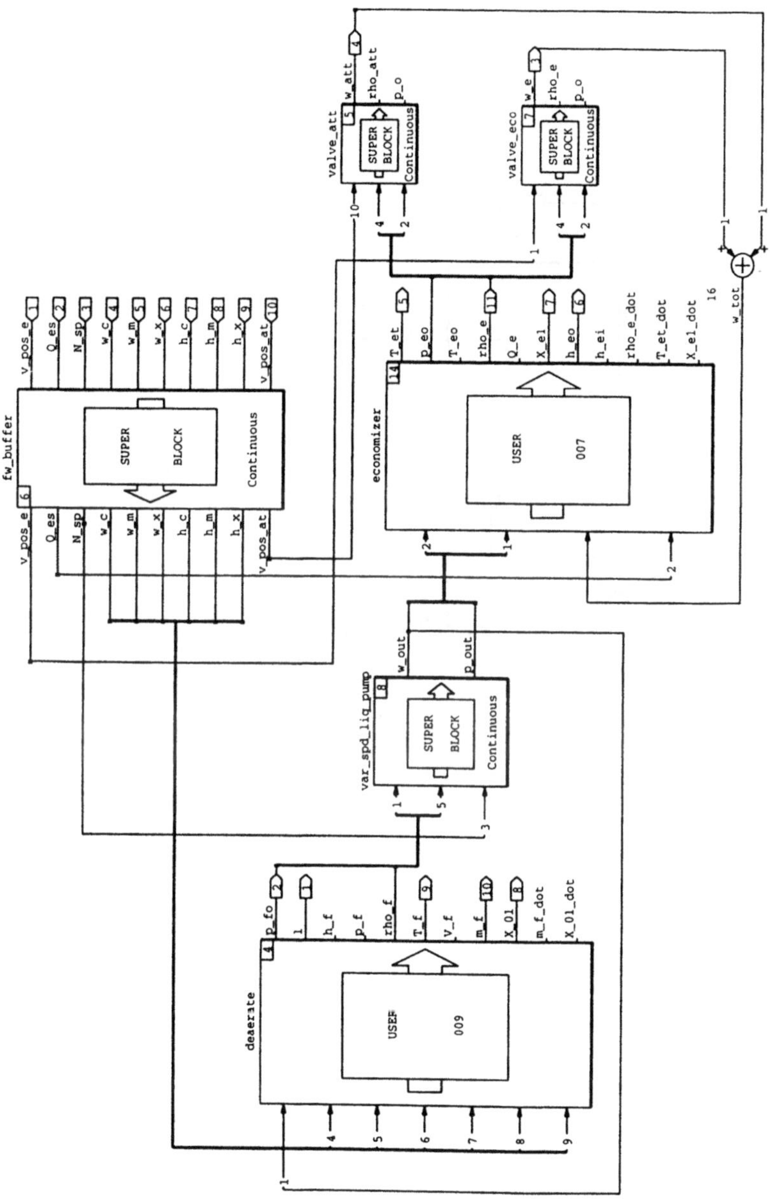

Fig. 5.13 Superblock FEEDWATER - internal structure.

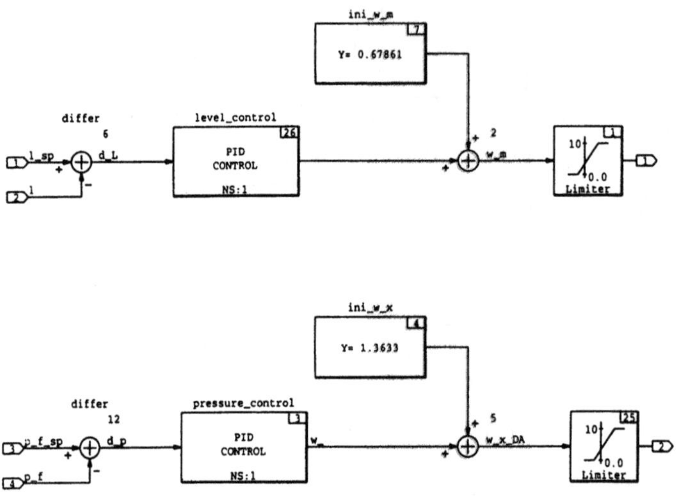

Fig.5.14 Feedwater system controller.

5.2.6 The Electrical Generator

The generator is constructed from the integration of the three superblocks defined using standard MATRIX$_X$ library blocks

- - SI_perunit_conversion_ST
- - generator_ST

- \- perunit_SI_conversion

Module **Generator** includes standard $MATRIX_X$ blocks. It is a simple linear model of the electro-mechanical relationships for a generator. All module internal variables are expressed in the per-unit system. The conversion blocks consist of simple gains to enable conversion of units between SI and the per-unit system. The overall structure of the generator model is depicted in Fig. 5.15.

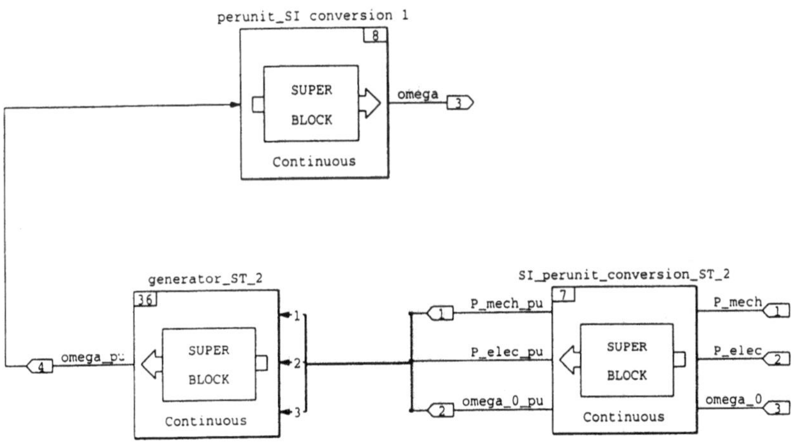

Fig. 5.15 The electrical generator.

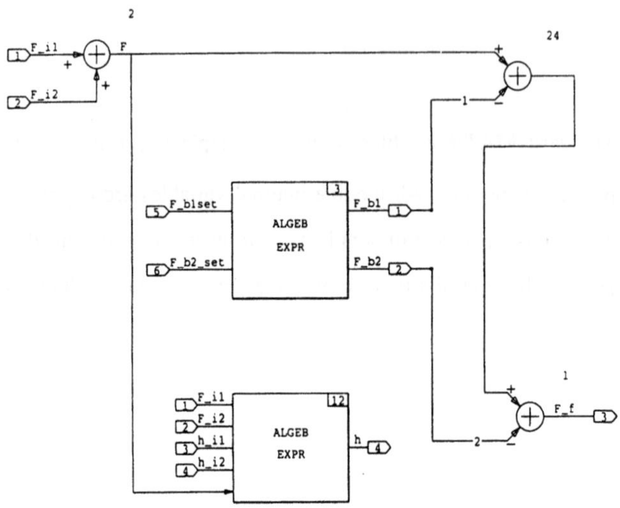

Fig. 5.16 Gas merge and splitting process.

5.2.7 Gas Merge And Splitting Process

The gas-merge and splitting process is constructed using standard $MATRIX_X$ library blocks. The process is shown in Fig. 5.16.

5.2.8 The Valve (Gas)

The valve (for gas flows) is constructed from the integration of the following modules:

- FORTRAN subroutine *Valve-gas.f*
- One superblock defined using standard $MATRIX_X$ library blocks *Servo*

The overall model structure is shown in Fig. 5.17.

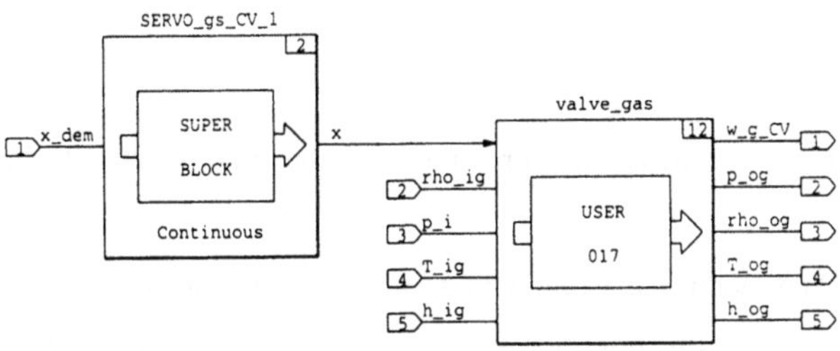

Fig. 5.17 Valve (gas)

5.3 Skegton Unit Simulation Results

5.3.1 Introduction

The distinguishing feature of combined cycle power (CC) plant is the joint production of electricity from a gas turbine and steam turbine, where the high heat content of the gas turbine exhaust flow is utilised to generate additional electricity by passing it through a waste heat boiler, that raises steam for admission to the steam turbine.

In the case of combined heat and power (CHP) plant, both electrical power and heat represent the delivered outputs. The heat energy may be in the form of hot water and used in district heating schemes, alternatively it may be in the form of steam and used for industrial processes such as drying.

In Chapter 1, the detailed description of all modules present in CC and CHP plant and the many different possible configurations has been presented. Furthermore a generic plant structure has been established for a dynamic simulation case-study. This configuration has been given the name Skegton unit. The primary objectives of the Skegton Unit design were as follows:

- To possess a generic plant structure capable of representing all the basic CC/CHP plant structures found in industry.
- To contain all the main power plant modules existing within CC/CHP power plants.
- To facilitate testing of all effects that are important when designing supervisory CC/CHP power plant control architecture.

As a result, Skegton Unit has been defined to comprise two power trains, each being a full CC/CHP configuration. Interconnections between the trains are possible, increasing the flexibility of the system. For example, exhaust gas from the gas turbine may be directed to the boiler in the same train or in the second train. A schematic diagram (flow diagram) of the Skegton Unit is presented in Fig. 1.23 in Chapter 1. The block diagram resulting from the $MATRIX_X$ implementation is depicted in Fig. 5.18.

In this section the tests applied to the Skegton unit to verify its correct operation are reported. Primary among the tests are the dynamic responses of controlled variables when a module reference set points are ramped or set to a new value. In addition open loop and closed-loop frequency responses are provided in the form of Bode diagrams. Finally, an example is provided on the application of the $MATRIX_X$ static model generation procedure from a given dynamic model.

Each power-train in the Skegton Unit consists of the following interconnected modules

- gas turbine
- boiler
- steam turbine
- condenser
- feedwater system

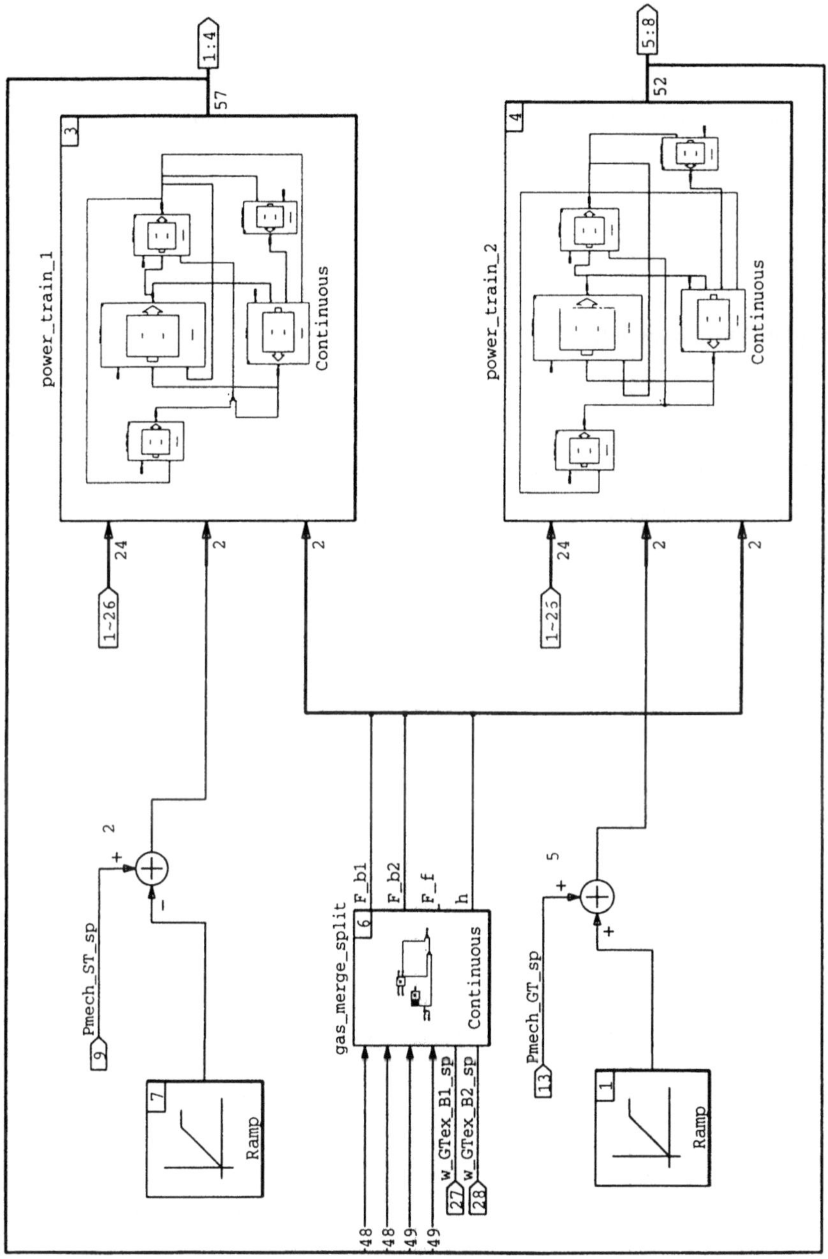

Fig. 5.18 Skegton unit - MATRIXx model.

A short description of each modules is presented in Section 5.2. Appendix 1 highlights inputs and primary outputs, and gives typical operating conditions for each module. For a detailed description of the mathematical modelling associated with each module please refer to Chapter 4 *Process models.*

5.3.2 Skegton Unit Sizing Data

5.3.2.1 Introduction

Although capable in principle of representing any particular CC/CHP configuration, the Skegton Unit as currently implemented is not based on an existing industrial installation. The availability of comprehensive CC/CHP plant operating data in the open literature is virtually nil. This is largely due to the commercially sensitive nature of such information. Consequently, for the first software version, the sizing of Skegton Unit has been based upon incomplete data. Detailed information for the boiler module was available from the references [*Chien et al., 1958; Nicholson, 1964*]. Thus other modules were sized to match the boiler.

The resulting fictitious CC/CHP plant may or may not be realistic in terms of engineering practicality, overall efficiency and size. Regardless of this, Version 1 of Skegton Unit satisfies the main design objectives and represent a valuable tool with which to conduct research into supervisory control strategies for CC/CHP plant.

5.3.2.2 An Approximate Steady State Operating Condition on Skegton Unit

An approximate steady state for the whole plant was established by performing a dynamic simulation of long duration (3 hours of simulation time) to allow all initial transients to reduce to zero. At this operating condition all the controlled outputs are settled at their reference set points. Considering the 33 dynamic states associated with a single power train's modules including the plant process states, excluding controller states, 29 of them exhibited a deviation of $< 0.1\%$ over a period of 10 minutes simulation time. There are sustained drifts in states associated with the boiler furnace (exhaust density and energy per unit volume) but these are limited to < 10% over 10 minutes. Note they are only indirectly influenced by low level controller action, eg control of furnace pressure. The approximate steady state operating condition is recorded in Appendix 1.

5.3.2.3 Sizing Data for Version 1 of Skegton Unit

Corresponding to the steady state operating condition presented in Appendix 1, the Skegton Unit salient full load operating data are as follows:

Gas Turbine

- Output power 34 [MW}
- exhaust gas flow 47 [kg/s]
- Exhaust gas temperature 1016 [^{O}K]
- Compression ratio 10 : 1

Boiler (waste heat with supplementary firing)

- superheated steam pressure 45 [bar]
- superheated steam temperature 717 [^{O}K]
- superheater steam flow 12 [kg/s]
- reheated steam pressure 13 [bar]
- reheated steam temperature 727 [^{O}K]
- Furnace fuel flow (supplementary firing) 14 [kg/s]

Steam turbine

- total output power 11 [MW]
- extraction steam flow 1.4 [kg/s]
- HP section outlet pressure 14 [bar]
- HP section outlet temperature 602 [^{O}K]
- HP section output power 3.4 [MW]
- IP section outlet pressure 5 [bar]
- IP section outlet temperature 610 [^{O}K]

- IP section output power 2.7 [MW]
- LP section outlet pressure 371 [mbar]
- LP section outlet temperature 376 [^{o}K]
- LP section output power 4.9 [MW]

Condenser

- operating pressure 60 [mbar]
- condensate flow 10.5 [kg/s]
- condensate temperature 309 [^{o}K]

Feedwater System

- deaearator operating pressure 640 [mbar]
- economiser outlet water flow 12 [kg/s]
- ecoomiser outlet water temperature 409 [^{o}K]

Skegton Unit

- Total output power = 2 x 34 [MW]
  + 2 x 11 [MW]
  = 90 [MW]

Note that for Version 1 of Skegton Unit there is no export of steam to outside users. The flow here represents that required internal to the Skegton Unit for injection to the gas turbine and the deaerator pressure regulation. This is taken after the HP section of the turbine.

5.3.3 Skegton Unit simulation data

The important data relevant to Version 1 of the Skegton Unit simulation are presented below.

- No. of Inputs 28
- No. of Outputs as required
- No. of States 132
- No. of Parameters ~ 200
- No. of FORTRAN subroutines 20
- No. of $MATRIX_X$ superblocks 20
- No. of hierarchical levels 6
- Integration Algorithm Variable-Step Kutta Merson
- Time step : determined by algorithm to satisfy specified error tolerance, maximum value of time step is 0.1 second
- No. of algebraic equations in FORTRAN subroutines : ~400
- No. of SISO low level control loops : 11 per power train
- Speed of simulation : 4× slower than real-time

5.3.4 Skegton Unit Dynamic Response Tests

5.3.4.1 Introduction

For each constituent module and for the Skegton Unit overall, the performance of the low level controllers when the set points are ramped or set to new values has been tested. It should be noted that the process of tuning the various PID control loops was only carried out approximately. Multiloop tuning of PID control to optimise performance (e.g tracking set point changes, rejecting disturbances, minimising loop interactions) remains an active area of research. The references [*Maciejowski 1989; Grimble and Johnson, 1988; Grimble, 1994*] presents various control design techniques for application to linear systems. For nonlinear systems, as is the case of power generation plant, linear analysis is only valid for small perturbations about a given operating condition and controllers may need to be made adaptive to achieve desired performance over the whole operating envelope. Often, in industry loops are tuned manually on a trial and error basis. The approximate tuning of the PID control loops will in some cases have a detrimental affect on the dynamic responses due to a step or ramp input. For example undesired peak overshoot or oscillatory behaviour may be present which improved tuning would remove.

For all tests it is observed that the controlled outputs attain their reference set point values after a transient period.

5.3.4.2 The Gas Turbine Module tests

- Test 1 + 10% ramp in demanded power output of duration 20 seconds
- Test 2 : + 10% step in demanded exhaust temperature

The dynamic responses of the controlled outputs and the manipulated input are shown in Figs. 5.19 -5.22

5.3.4.3 The Boiler Module test

- Test 1 + 10% step in drum level set point
- Test 2 : - 5% step in superheat steam temperature set point

The dynamic responses of the controlled outputs and the manipulated input are shown in Figs. 5.23-5.26.

5.3.4.4 The Steam Turbine Module test

- Test 1 - 10% ramped in demanded power output of duration 20 seconds.

The dynamic responses of the controlled outputs and the manipulated input are shown in Fig. 5.27.

5.3.4.5 The Condenser Module test

- Test 1 - 2% step in condensate temperature set point

The dynamic responses of the controlled outputs and the manipulated input are shown in Fig. 5.28.

5.3.4.6 The Feedwater Module test

- Test 1 + 10% step in deaerator level set point

The dynamic responses of the controlled outputs and the manipulated input are shown in Fig. 5.29.

5.3.4.7 The Skegton Unit test

- Test 1 Demanded power output from the steam turbine in power train 2 is ramped by 10% whilst simultaneously demanded power output from the gas turbine in power train 2 is ramped by + 10%.

The dynamic responses of the controlled outputs and the manipulated input are shown in Figs. 5.30-5.33.

5.3.5 An example of using MATRIXx to determine a static model

Given the dynamic model of a power plant or of constituent modules such as the boiler, steam turbine etc., it is often required to establish the equivalent static model for given operating point.

In $MATRIX_X$ an algorithm called *eqpoint* is used to achieve the above. Here, the algorithm is applied at a higher level in the Skegton Unit hierarchy to the condenser module. The condenser module contains four FORTRAN subroutines representing the shell, tube, steam and water equations respectively, and the condenser controller. The three steps taken in applying the eqpoint procedure successfully to the condenser module are summarised below.

- An initial FEASIBLE operating point for the model must be located. This may be done by considering the inherent physical constraints present in the normal operation.In the case of the condenser this means the following :

(i) The tube liquid temperature must increase from section to section along the vertical axis of the condenser between inlet and outlet

(ii) The temperature of the tube and shell metal must be less than or equal to the condensing steam temperature

(iii) All flows of temperatures and enthalpies are positive quantities.

- The routine eqpoint may then be applied with the start point given by an initial feasible operating point. The static model for given inputs, states or outputs existing at the feasible start point is then located by an iterative Newton-Raphson root finding procedure. For the condenser the inputs at the feasible start point were fixed (frozen) and the states and outputs were allowed to float, representing degrees of freedom for the eqpoint algorithm.
- Having located a static model, it may be required to establish the new static model for one or more inputs, states or outputs changed. For the condenser example, one of the inputs, the inlet steam flow is reduced by 20% from 12.5 [kg/s] to 10 [kg/s] and the new static model for fixed inputs is computed.

Fig. 5.34 shows the dynamic responses of the condenser model when initialised to a feasible start point. Note that this start point is not close to a steady state operating condition. However after applying eqpoint, the perfect static model is achieved after only 10 iterations of the algorithm. The resulting dynamic responses confirm that the located operating point is indeed a steady state, Fig. 5.35.

5.3.6 Selected Frequency Responses for Skegton Unit - Bode Diagrams

Frequency responses diagrams can be utilised in for example low level controller design or model simplification. They provide an indication of the degree of stability (via Nyquists criterion) and also of performance characteristics such as speed of response or bandwidth sensitivity to noise.

For the Skegton Unit, as previously intimated, the low level control design problem is a complex one, since the unit is nonlinear with multiple control loops. Thus no attempt will be made here to interpret the Skegton Unit Bode diagrams for low level controlled design.
Some selected frequency responses associated with the gas turbine, boiler and steam turbine modules are presented in Figs. 5.36-5.38.

5.4 Conclusions

The main objective of the Chapter was to verify the Skegton Unit's global operation. A short description of the simulation modules has been given. Dynamic responses of controlled outputs to step and ramp changes in the reference values have been presented. In addition, frequency responses for selected modules have been calculated. The successful application of the $MATRIX_X$ static model finder algorithm to the condenser module has been demonstrated. A summary of Skegton unit modules sizing data and simulation data is also included.

5.5 References for the Chapter

Chien K.L., E.I. Ergin, C. Ling and Allyn Lee ***Dynamic analysis of a boiler.*** Transactions of the ASME, November 1958

Grimble M.J. ***Robust industrial control. Optimal design approach for polynomial systems.*** Prentice Hall, 1994

Grimble M.J. and M.A. Johnson ***Optimal control and stochastic estimation.*** John Wiley and Sons, 1988

Maciejowski J.M. ***Multivariable feedback design.*** Addison Wesley, 1989

Nicholson H. ***Dynamic optimisation of a boiler.*** Proc. IEE Vol.111, No.8, August 1964

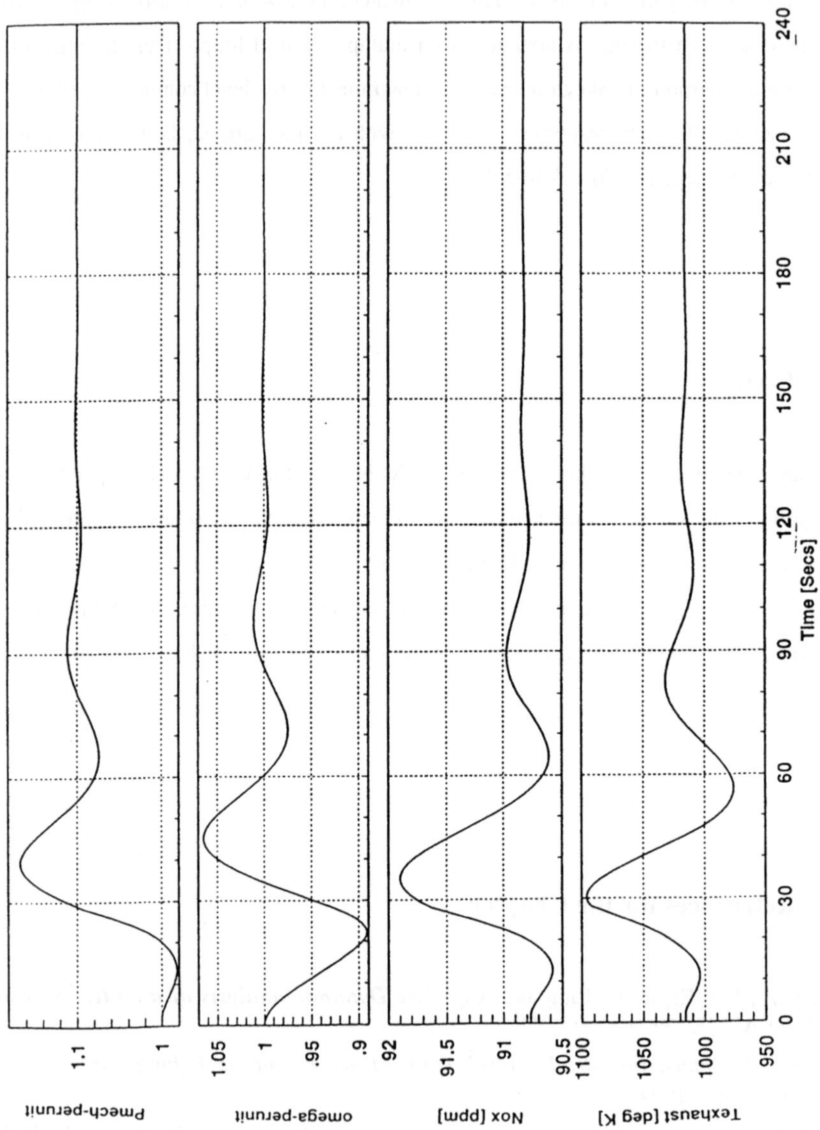

Fig. 5.19 Gas turbine module test results. *+10% Pmech_sp & Pelec ramp change of duration 20 seconds - controlled outputs.*

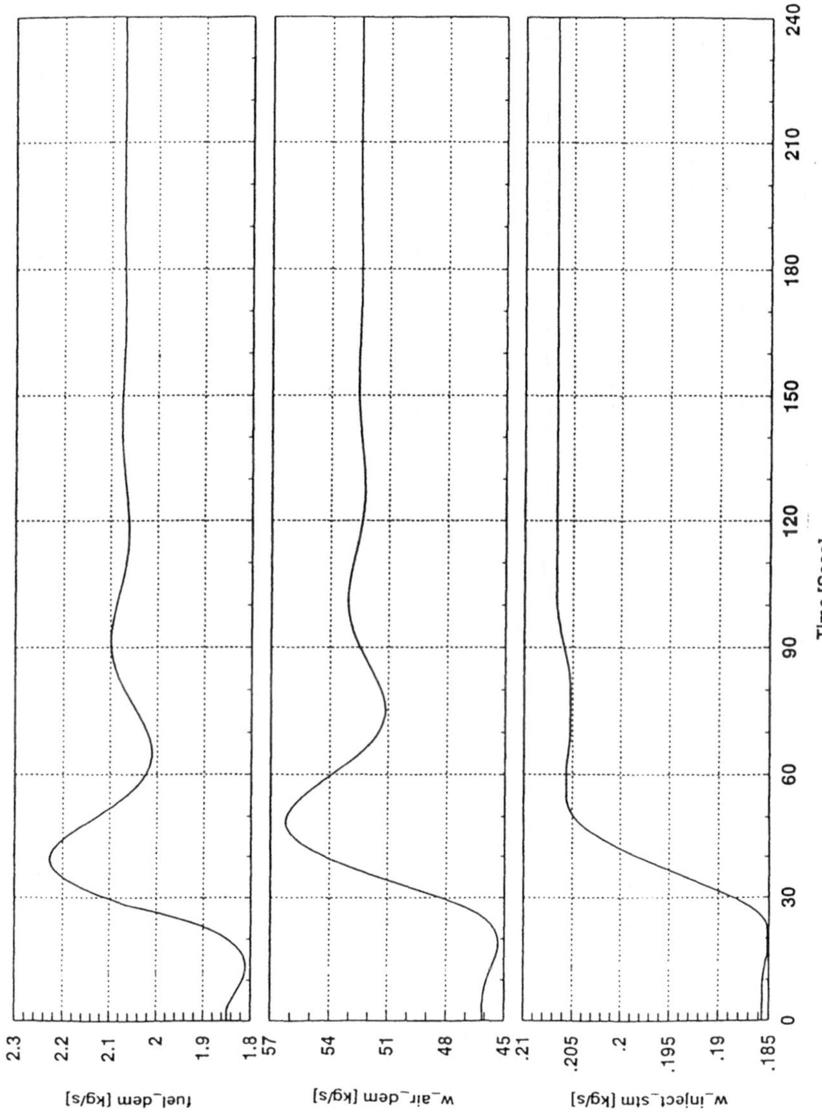

Fig. 5.20 Gas turbine module test results. *+10% Pmech_sp & Pelec ramp change of duration 20 seconds - manipulated inputs.*

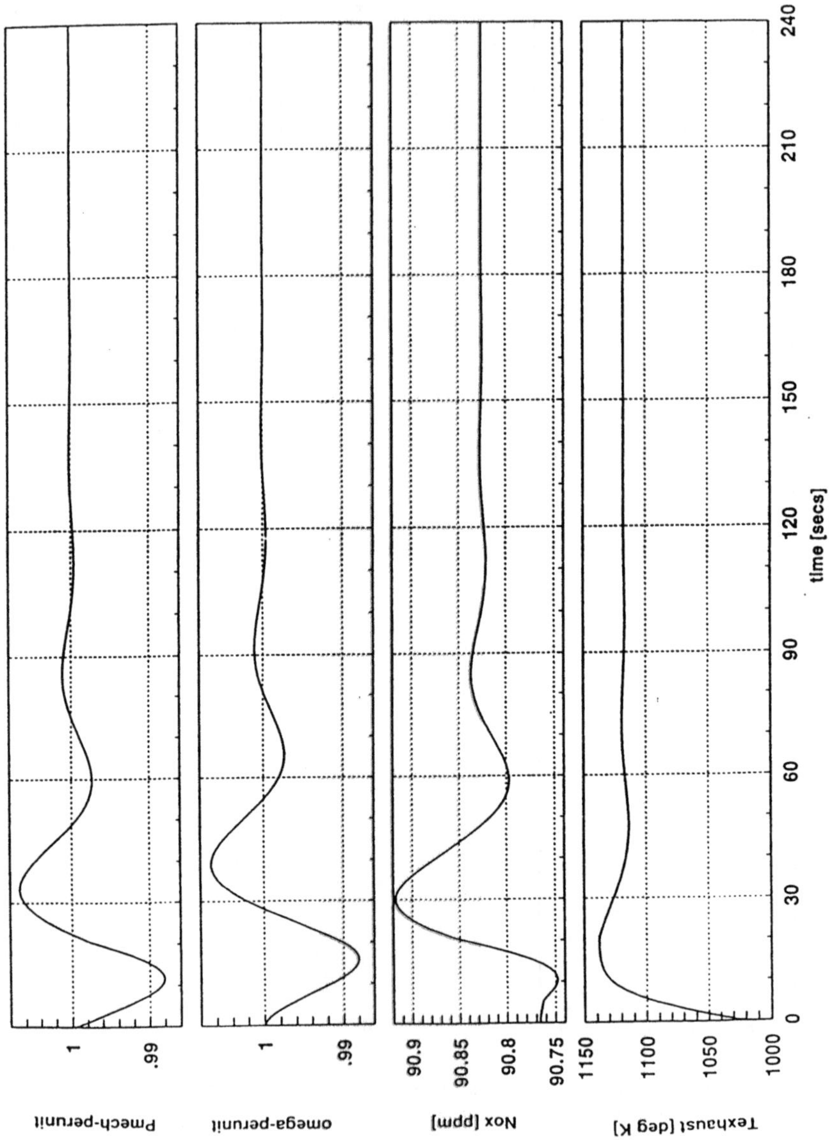

Fig. 5.21 Gas turbine module test results. *+10% step in Texhaust set-point - controlled outputs.*

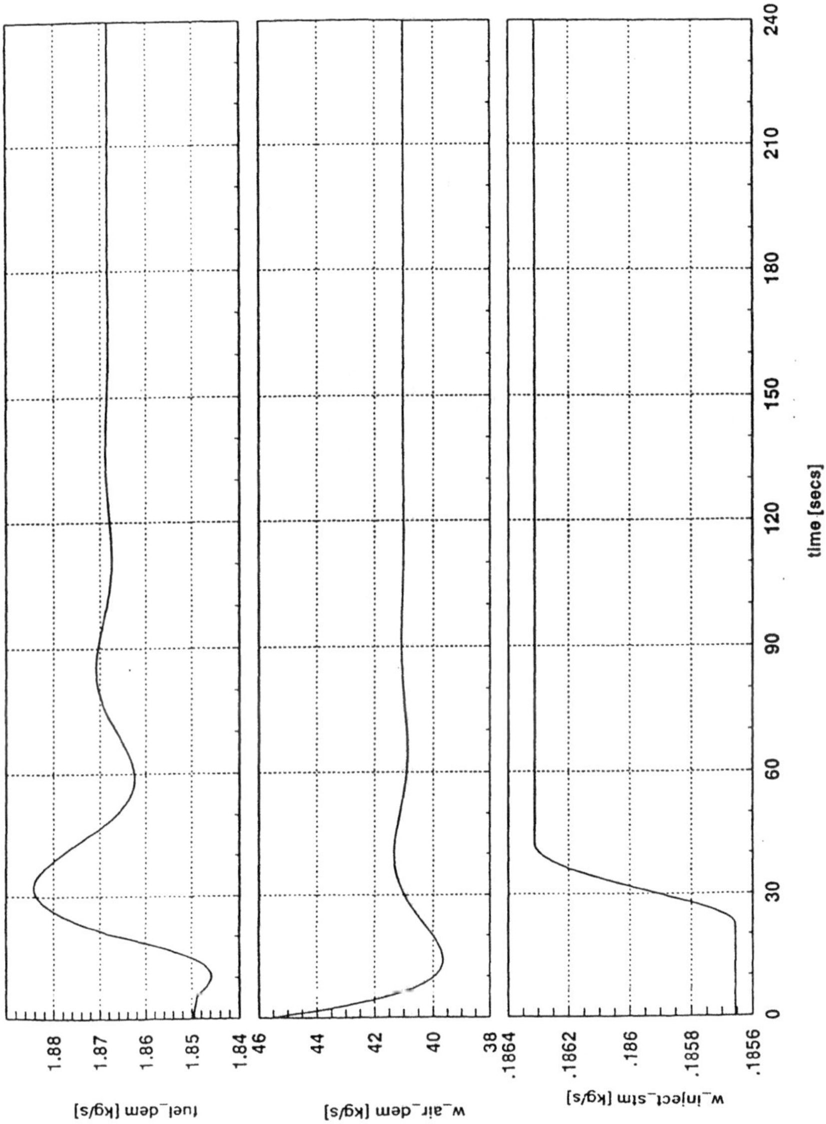

Fig. 5.22 Gas turbine module test results. *+10% step in Texhaust set-point - manipulated inputs.*

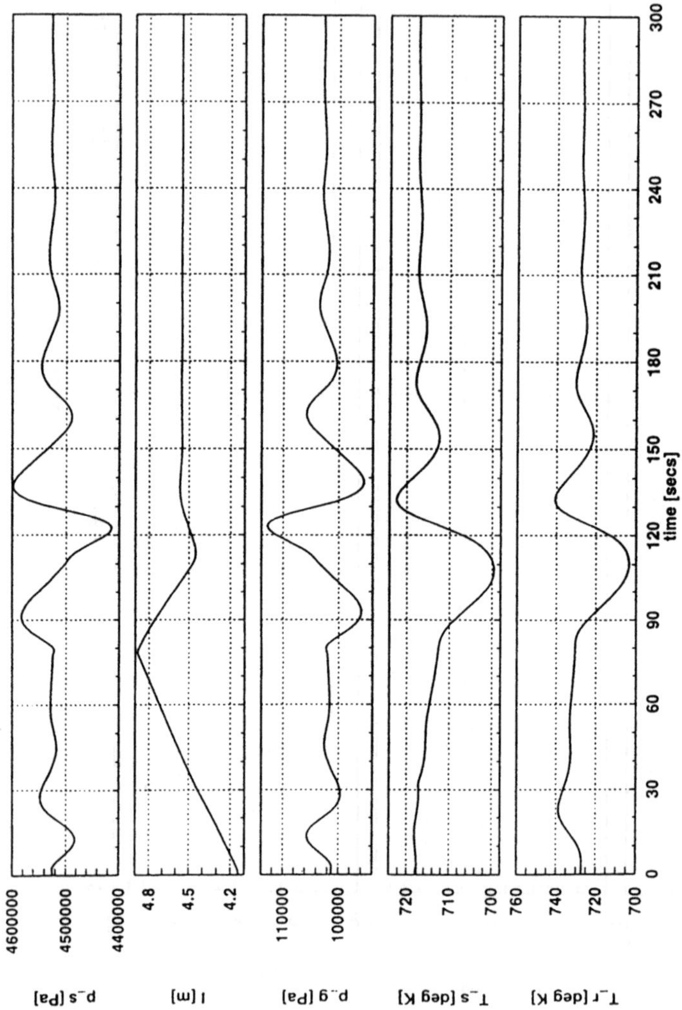

Fig. 5.23 Boiler module test results. *+10% step in drum level set point - controlled outputs.*

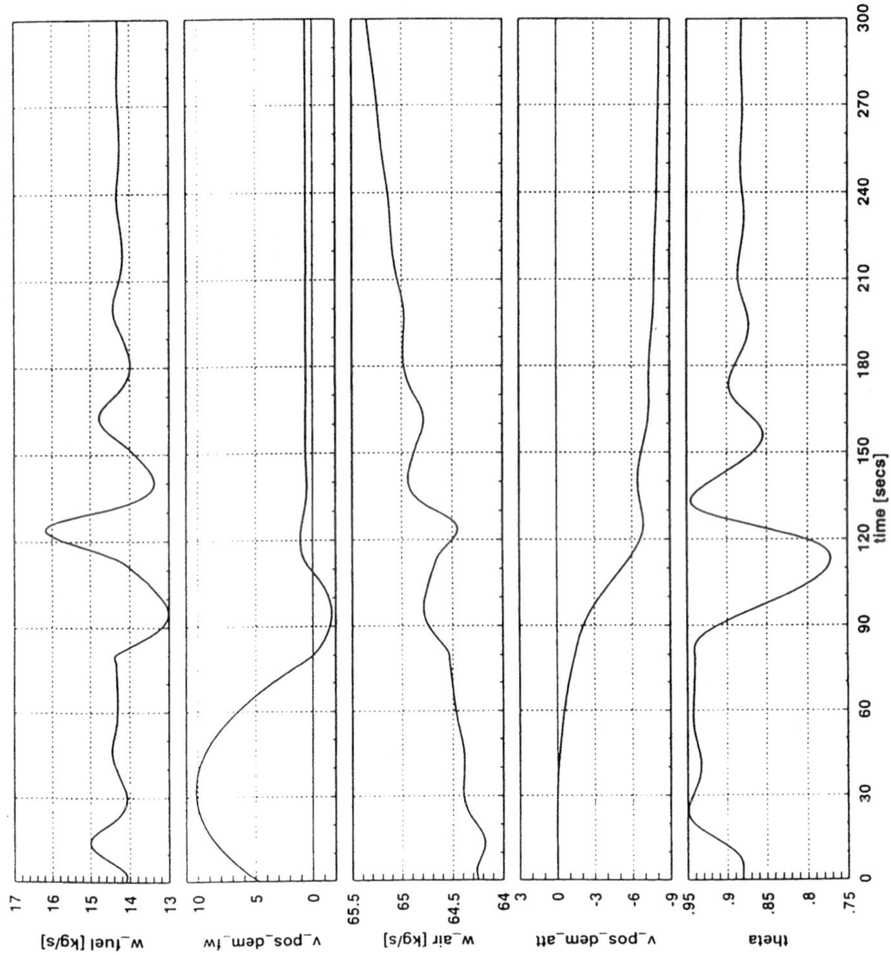

Fig. 5.24 Boiler module test results. *+10% step in drum level set point - manipulated inputs.*

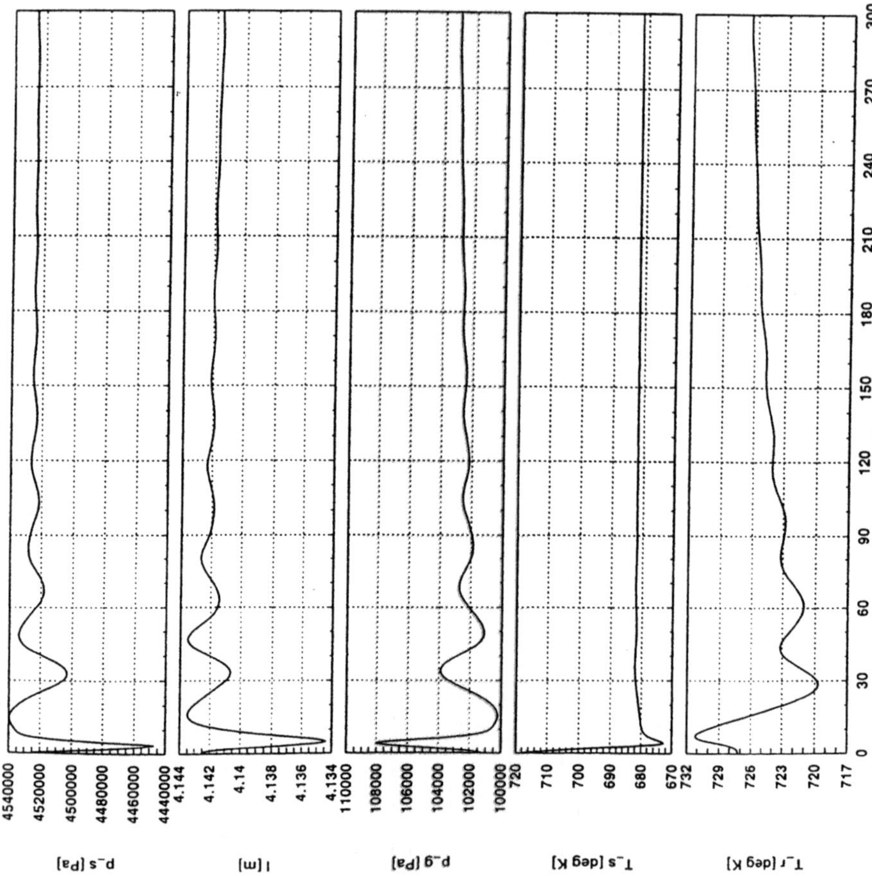

Fig. 5.25 Boiler module test results. *-5% step in superheat steam temperature set point - controlled outputs.*

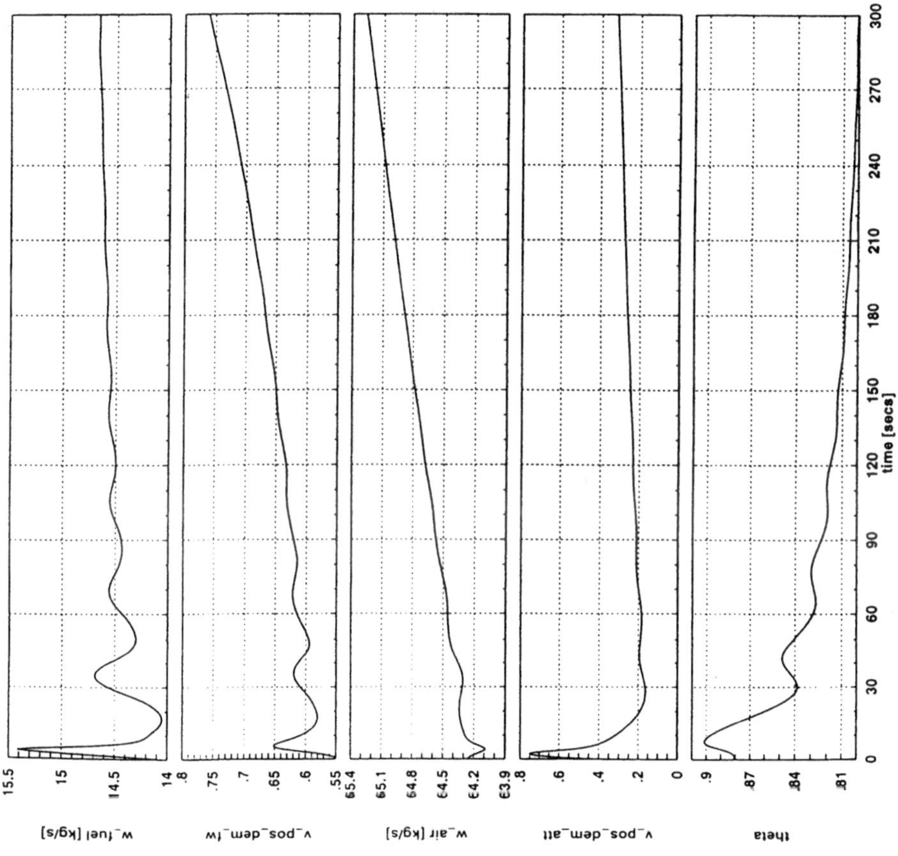

Fig. 5.26 Boiler module test results. *-5% step in superheat steam temperature set point - manipulated inputs.*

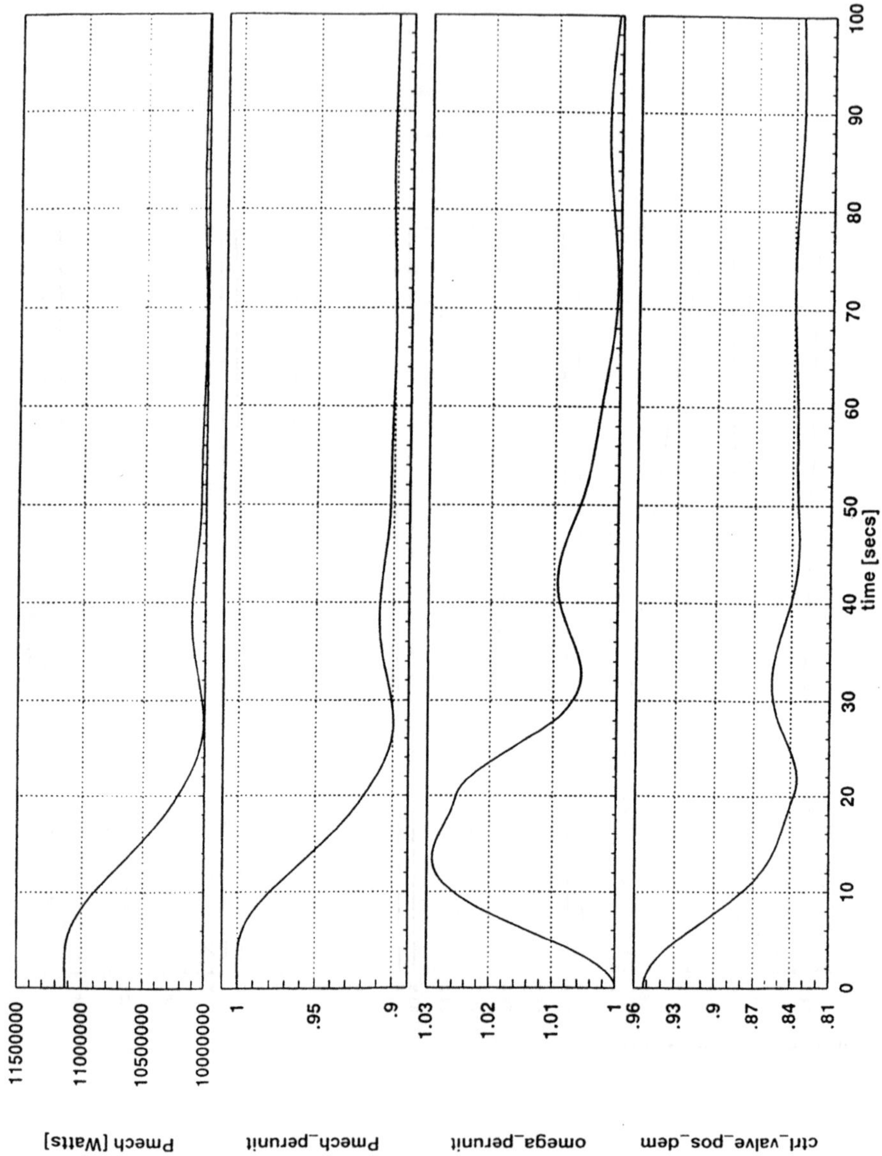

Fig. 5.27 Steam turbine module test results. *-10% Pmech_set_point & Pelec ramp change of duration 20 seconds.*

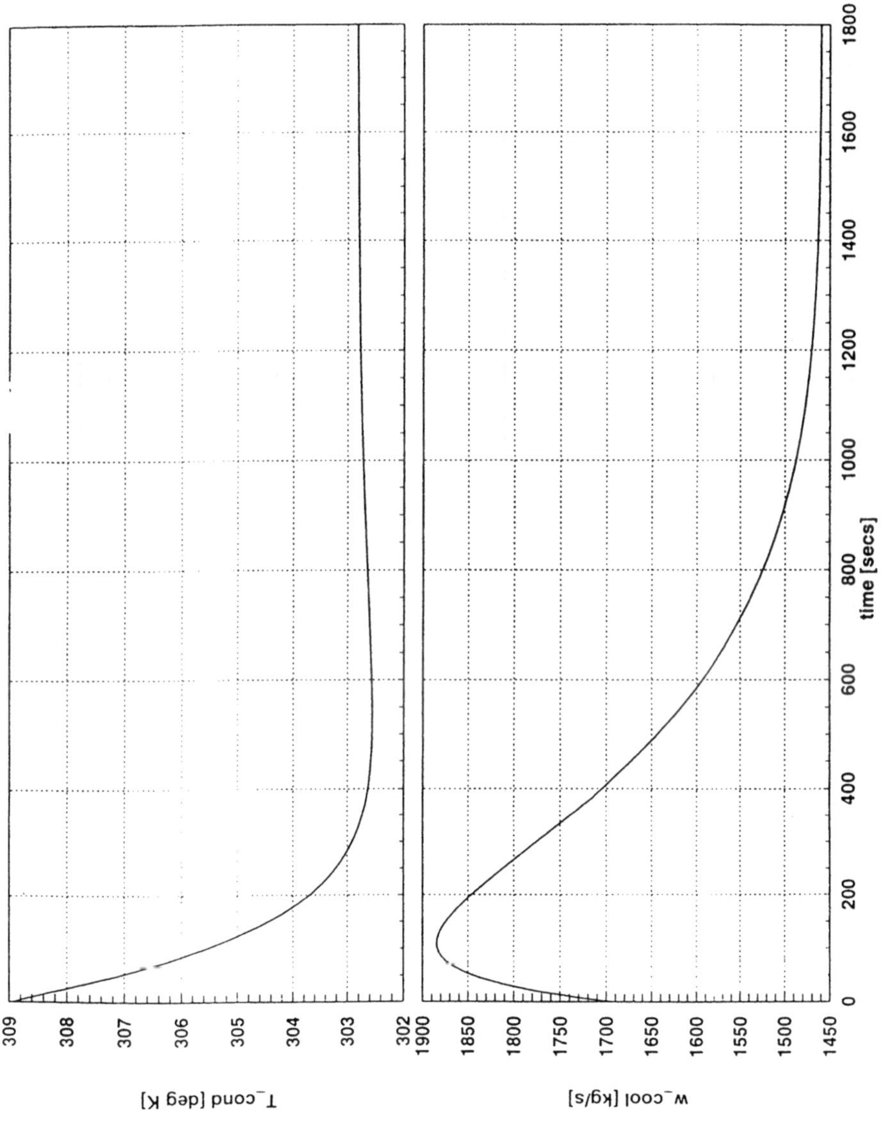

Fig. 5.28 Condenser module test results. *-2% step in condensate temperature set point.*

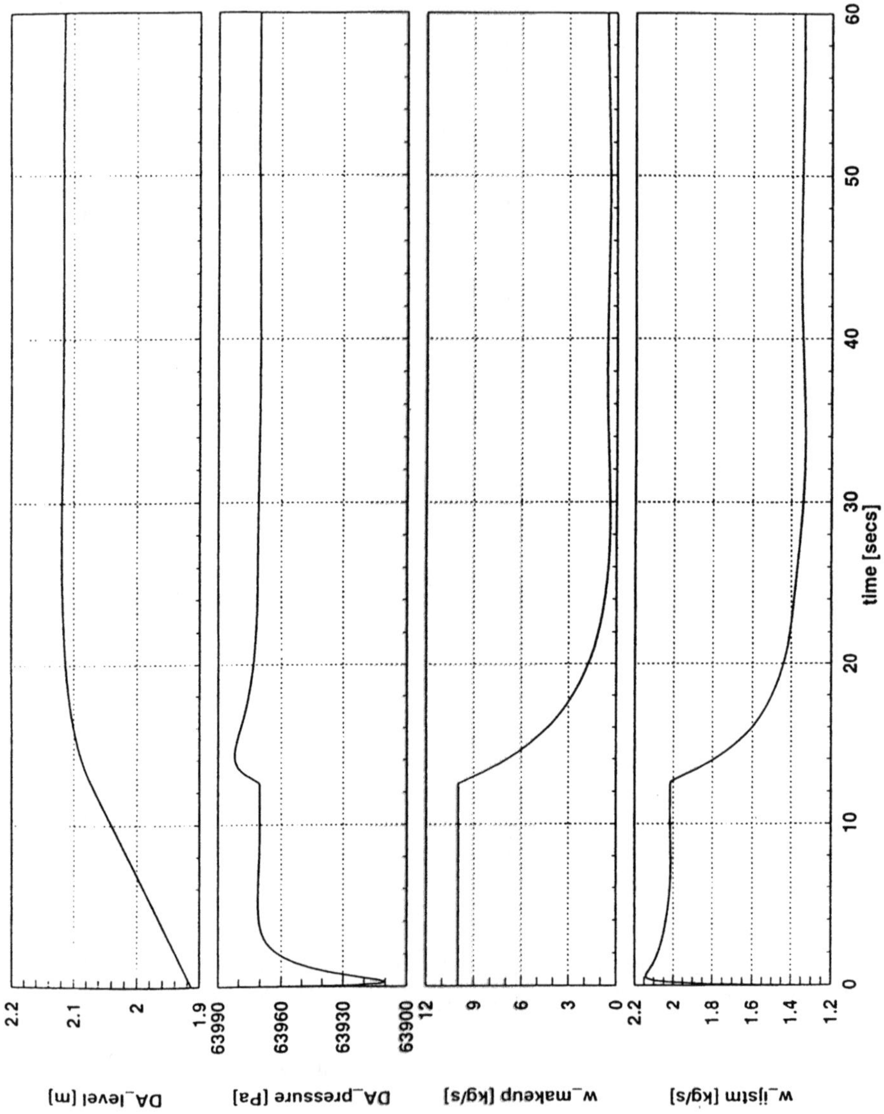

Fig. 5.29 Feedwater system module test results. *+10% step in deaerator liquid level set-point.*

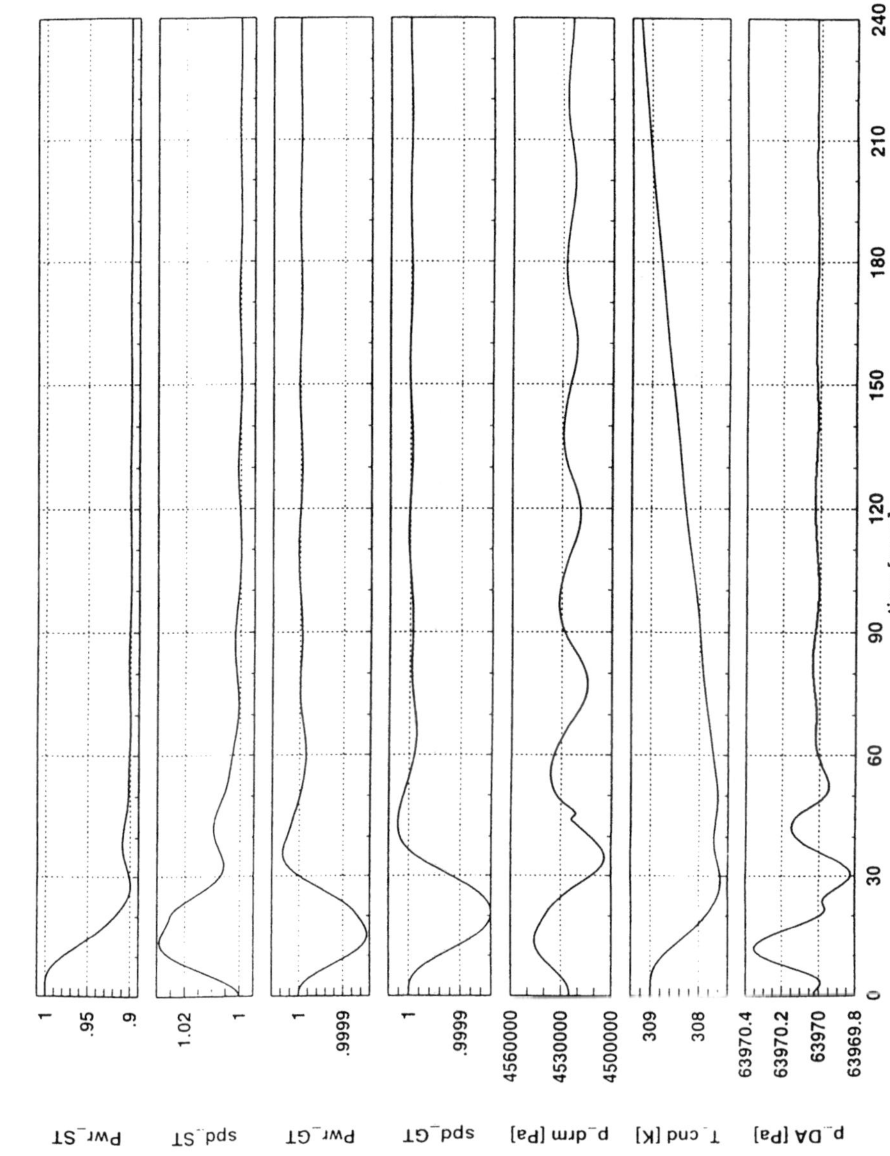

Fig. 5.30 Skegton Unit test results. *-10% ST1 & +10% GT2 set-points ramped; power train 1 - controlled outputs.*

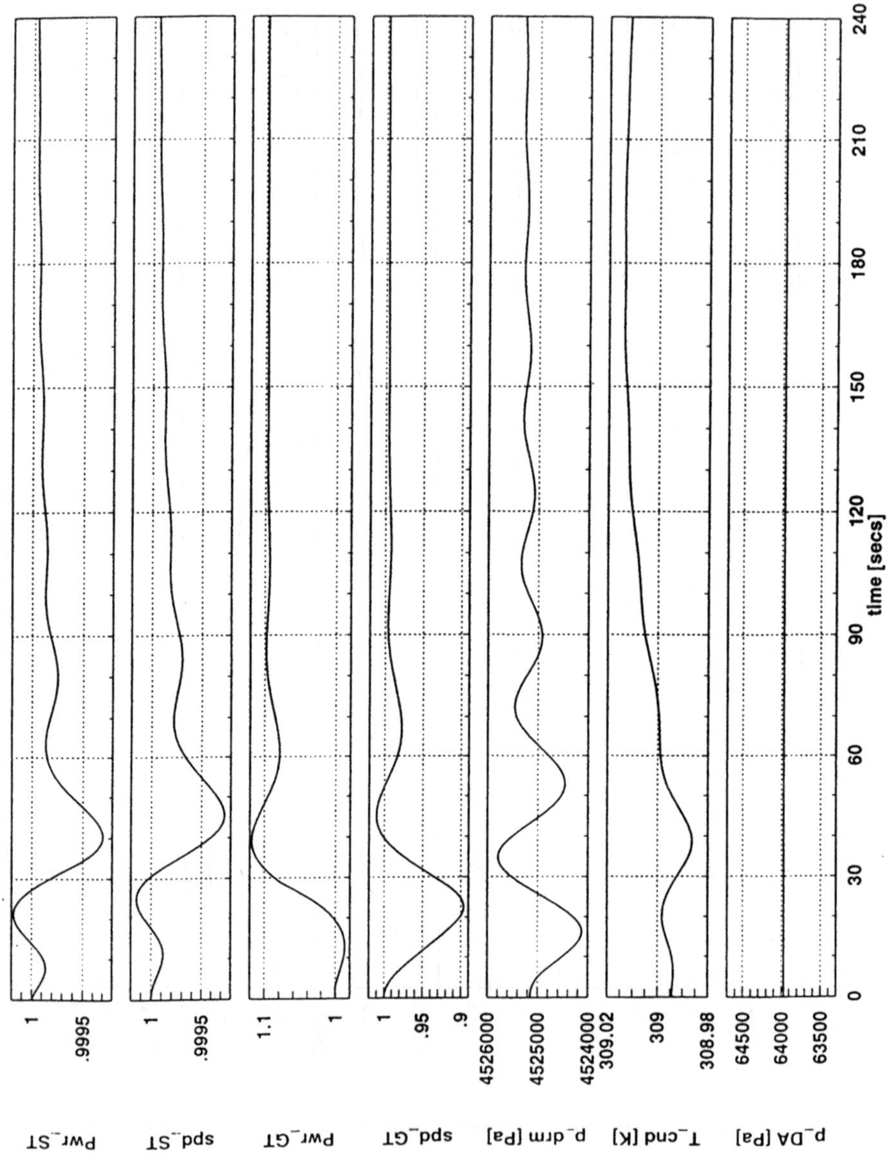

Fig. 5.31 Skegton Unit test results. *-10% ST1 & +10% GT2 set-points ramped; power train 2 - controlled outputs.*

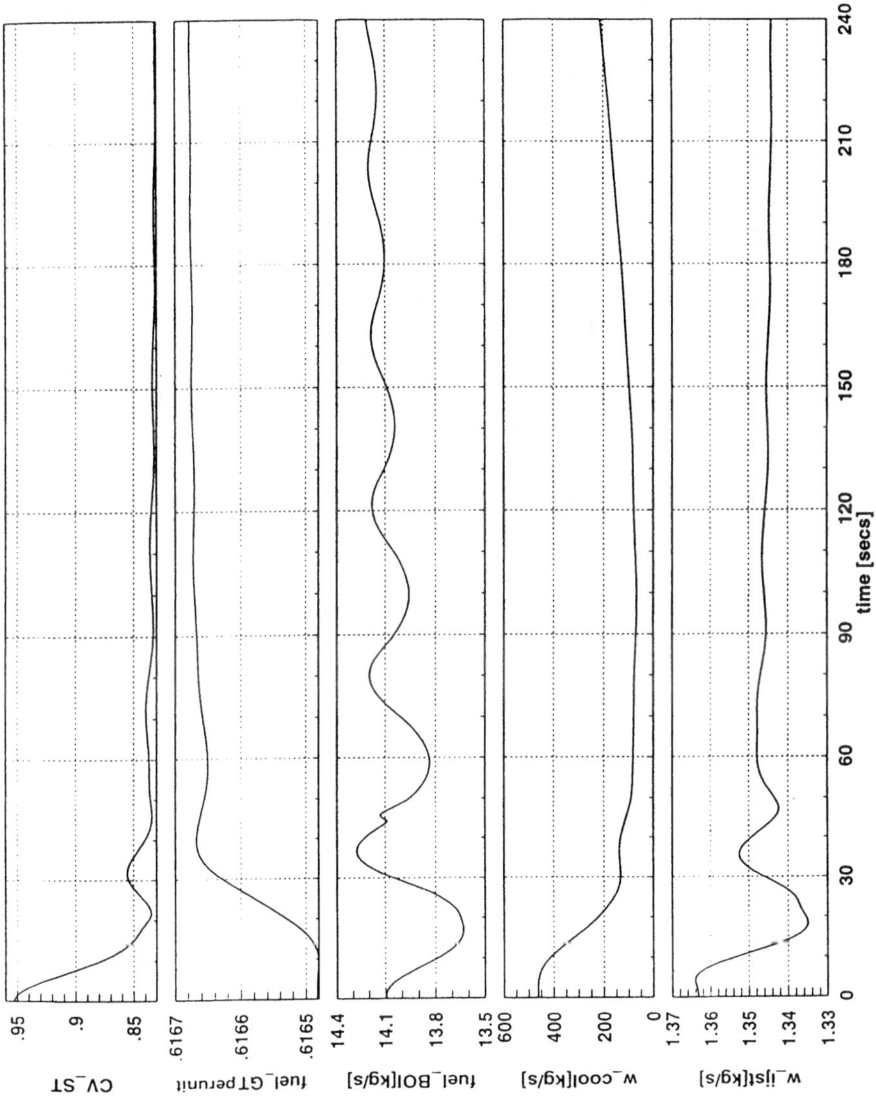

Fig. 5.32 Skegton Unit test results. *-10% ST1 & +10% GT2 set-points ramped; power train 1 - manipulated inputs.*

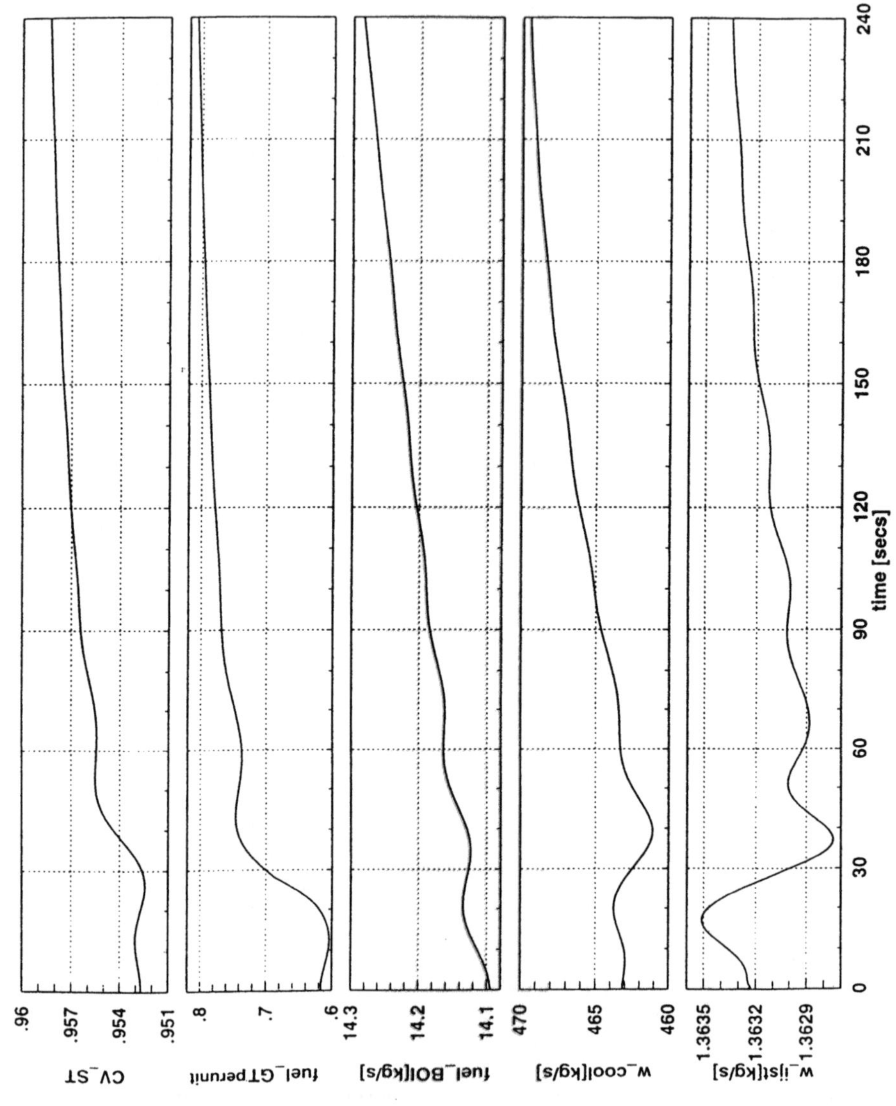

Fig. 5.33 Skegton Unit test results. *-10% ST1 & +10% GT2 set-points ramped; power train 2 - manipulated inputs.*

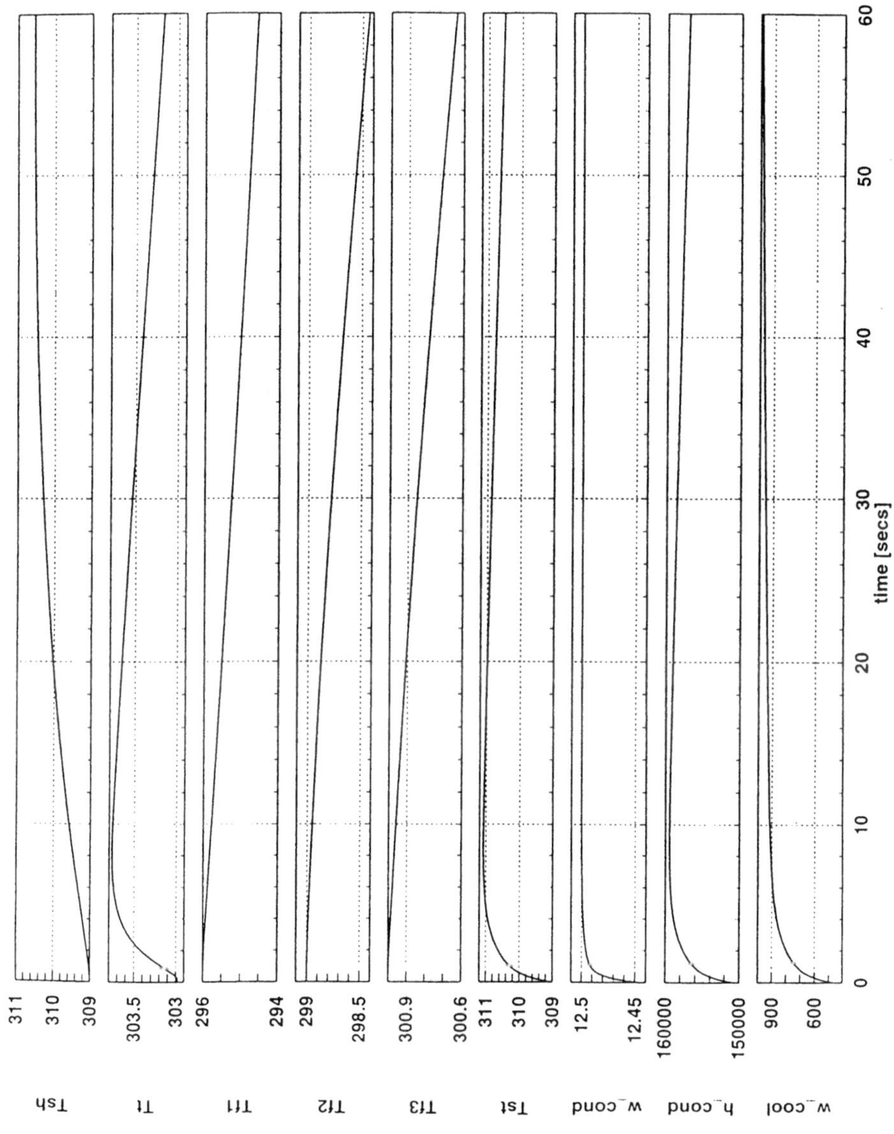

Fig. 5.34 Condenser module test results. *Model responses when initialised to a feasible operating point.*

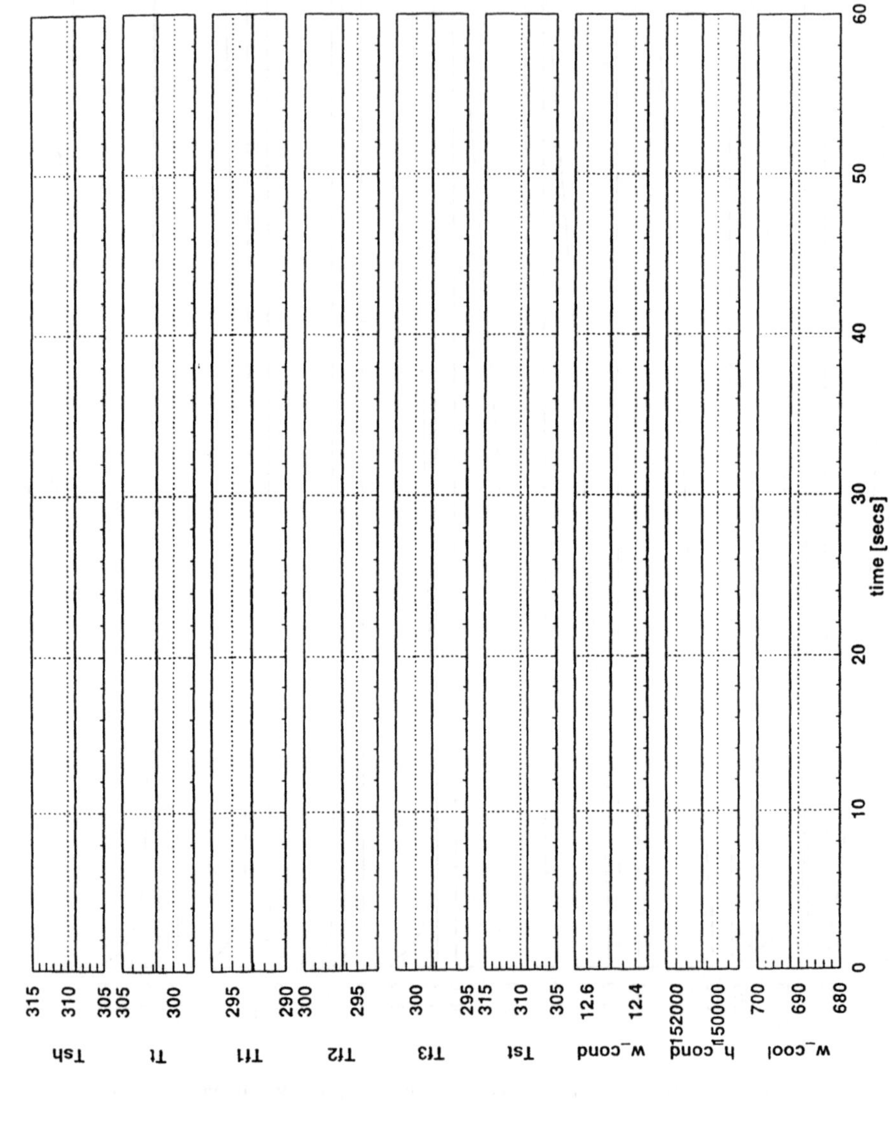

Fig. 5.35 Condenser module test results. *Model responses when initialised to a MATRIXx located steady-state operating point.*

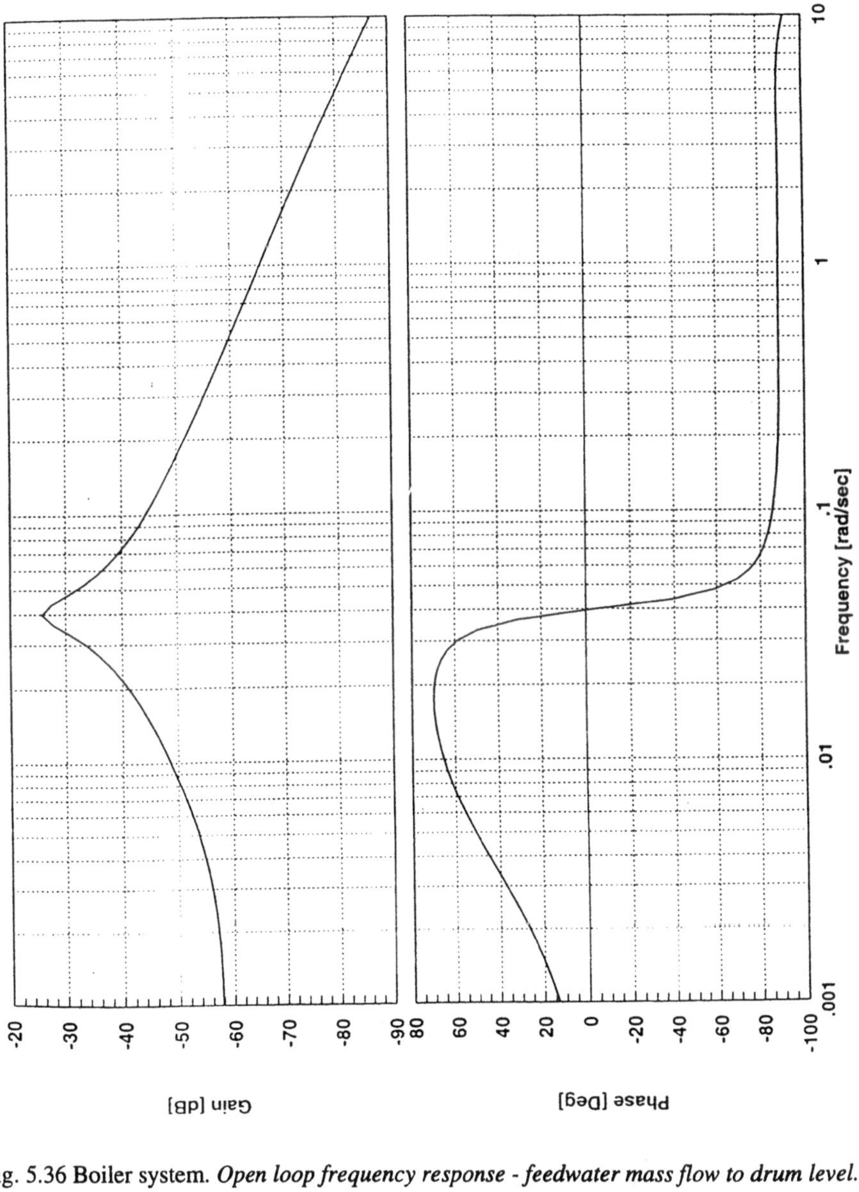

Fig. 5.36 Boiler system. *Open loop frequency response - feedwater mass flow to drum level.*

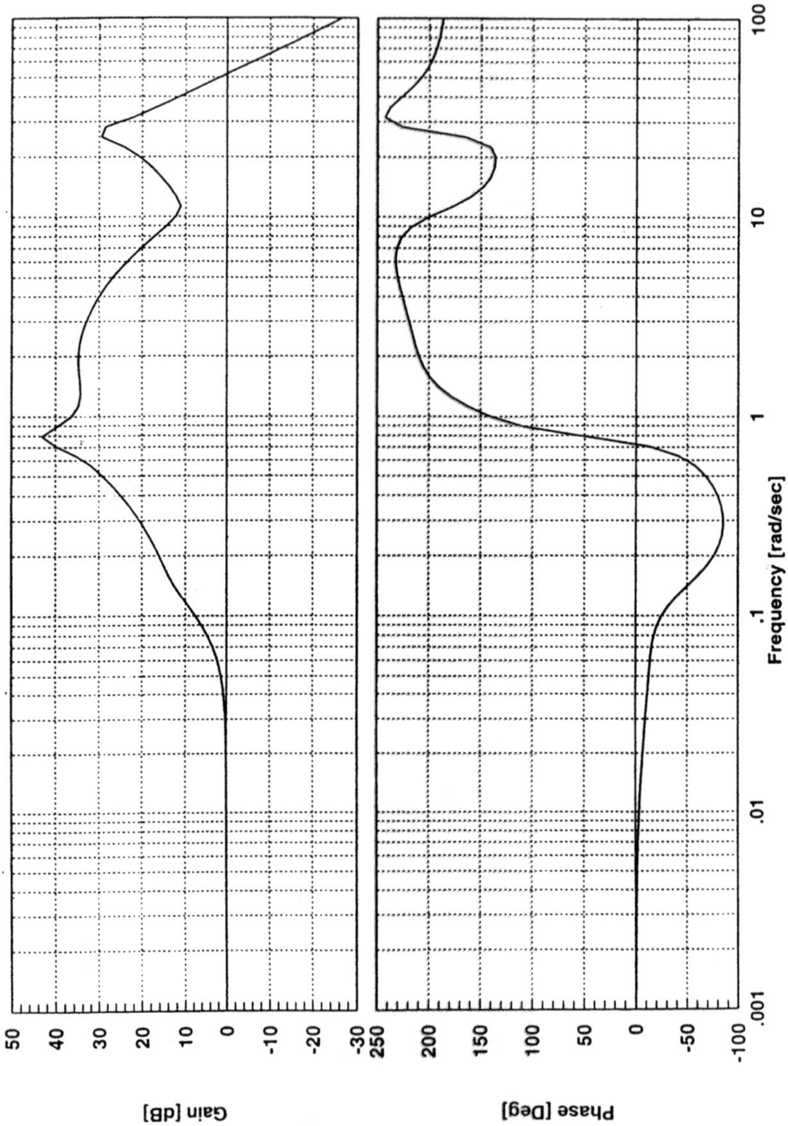

Fig. 5.37 Gas turbine system. *Closed loop frequency response - power set-point to power.*

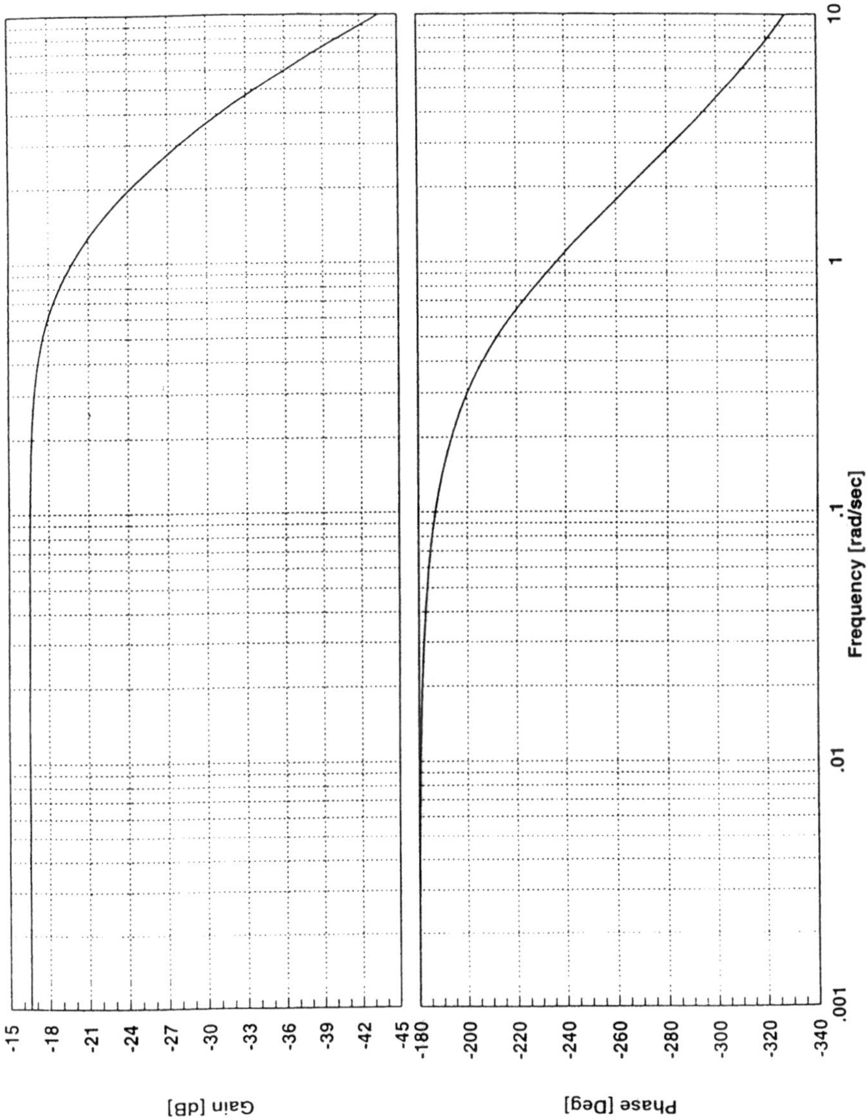

Fig. 5.38 Steam turbine system. *Open loop frequency response - control valve position to power.*

CHAPTER 6

CONCLUSIONS AND FUTURE DIRECTIONS

The outcome of the first phase of a power generation plant research programme is reported in this monograph. In this phase the dynamic simulation framework was established and some generic control research directions identified. It is useful to conclude by recording the main findings of the study and preview briefly the possible future directions.

6.1 Summary Conclusions

Most industrial systems have an associated pre-history of published papers and books describing the system technology and behaviour. Power generation plants are no exception and the available open literature formed the main source of information for the research reported. This was commented on by industrial experts to ensure compatibility with current technology and operational practice. The objective of this literature search was to identify the list of basic components from which to construct standard and non-standard CC and CHP configurations. The basic configuration and component subsystems were captured using the straightforward input-output block diagram schematic. It is our past experience that process engineers are not so familiar with this schematic since they prefer to work with flowsheets or the Piping and Instrumentation Diagram (P and ID). For the control systems engineer, the block diagram brings clarity to the system structure by emphasizing the important connections and interactions for understanding the system behaviour. Two other aspects of the utility of this formalism are that block diagrams enable new configurations to be easily assembled and that the software tools to be used are based on the hierarchical block diagram structure and set-up. The whole of the first chapter is directed towards extracting from highly complex power generation plant configurations, the essential input and output structures necessary for system simulation and control. This work on CC/CHP configuration and structure closes with the specifications of Skegton Unit. Skegton unit was devised as a test plant for the ideas of the subsequent research. It is a fairly complicated power generation plant which was used to inspire the research on system simulation, regulator loop design and supervisory control system concepts. Of course, Skegton unit is not a real power generation plant but its design was a very useful conceptual catalyst for the research.

The long term objective of the research is to establish a framework and portfolio of control techniques for the regulating loops, the supervisory control and the interaction between supervisor and regulator. The simulation library is seen as an essential item to achieve this aim. However, to ensure the immediate tasks of constructing a power generation plant simulation library are compatible with the longer term aims of the control research, a preliminary control review was performed. This is reported in Chapter Two. An appreciation of the current and future potential of distributed control system technology is an important factor in any supervisory system. Of particular interest were the advances in plant-wide communications since these will enable a more responsive and globally pro-active industrial plant. The middle sections of this chapter review the control tasks within power generation plant and the utility of the Skegton unit for different types of control studies. This review is given a more coherent framework in the last section of the chapter. Here a strong indication is given of how the control research tasks might develop. The simulation clearly has an important supporting role to play in these developments.

The art of system simulation has not changed so drastically over the last twenty years or so. The basic mathematical models describing physical processes only really change when the underlying process technology changes. However, the *tools* to accomplish system *simulation* have undergone a radical change in the last decade. Windows, the mouse and the GUI are the visual aspects of this technological advance. Thus, the opportunity to assess state of the art simulation tools was both timely and valuable. Chapter Three reports this assessment. The steps involved were straight-forward : a look at system types, a look at the simulation tools available in the marketplace, the construction of an assessment basis followed by a short-list of software for closer investigation. The second part of the exercise centred on a simple but useful benchmark example of a boiler model. Thus screen displays, graphical output, speed of acquiring the skills to use the software and other aspects could be experienced, debated and illustrated. The journal *Measurement and Control* kindly facilitated a wider exposure for this material in the July, 1994 issue.

The core of the Monograph lies in Chapter Four. Here can be found the basic library of component models which were subsequently implemented in software. The presentation is perhaps a little formal but then Chapter Four is the model archive. Once more recourse was made to the authors' past experience is establishing a library approach to large-scale system modelling and simulation. The steps may be tedious but they are careful, systematic and successful:

- System decomposition to identify physical subsystems.
- Literature review to avoid replicating existing modelling work.
- Model assumptions to set a clear framework for discussion and comprehension.
- Model equations establishing the necessary level of complexity in modelling system effects.
- Block diagrams indicating a clear input-output system specification.
- Algorithm of the model to be used as a framework for implementation

Two principles are inherent in this approach. Firstly between model assumptions and model equations is the need to trade model complexity, model accuracy and computational load to arrive at a satisfactory model. This is the modeller's art. Secondly, whilst there is one algorithm for a model there are many implementations. This is the ideal situation where the algorithm embodies the essential steps in equation manipulation and the implementation is just one practitioners program of these essential steps. Striving to attain this ideal separation, helps to clarify and shorten the eventual implementation.

The final steps in this long process (in this case taking one year and four months) are the simulation trials, and model verification tests. The results for this process are reported in Chapter Five. The fact that the system is decomposed into subsystems and modules facilitates a comprehensive sequence of increasingly complex verification tests. These can mirror real plant acceptance tests if necessary. The work reported here archives single module tests (transient and steady-state accuracy assessments) and a test for the composite power generation plant configuration of Skegton unit. Of course in practice, real data should be used to validate the model.

6.2. The Future Research Directions

The foresight review of Chapter Two coupled with the generic framework used to develop the simulation software indicates that a possible outcome of the research could be a CC/CHP software toolbox. It has been given the tentative structure as shown below:

PROCESS LIBRARY	**DESIGN LIBRARY**
Boiler	Allocation Algorithm
Gas Turbine	Constrained Control
Steam Turbine	Loop Tuner
-----	----
-----	----
System Simulation	Control Algorithms

CC/CHP SYSTEM TOOLBOX

Typical endusers of such a toolbox could be industrial or academic. The industrial enduser might be able to use the software for in-house design studies, for training engineers or operators. For the academic, the package might be invaluable as a teaching tool on undergraduate or postgraduate courses for power systems engineering. For further information about the availability of the power system toolbox, the reader should contact either Dr. Johnson or Dr. Katebi at the Industrial Control Centre, University of Strathclyde.

However, the simulation development is an important step toward the control research. Thus far it has helped to identify the type of control problem which exists and it is hoped to report on these research developments in a subsequent publications.

APPENDIX 1

SKEGTON UNIT SIMULATION DATA

A1.1 Boiler

A1.1.1 The Furnace

Inputs :

- w_F = fuel flow to the furnace [kg/s]
- w_A = air flow to the furnace [kg/s]
- h_G = enhalpy of exhaust gas from the gas turbine [J/kg]
- w_G = exhaust gas flow from the gas turbine [kg/s]
- theta = tilt angle coefficient [-] (0<theta<1)
- T_st = temperature of superheater metal tubes [°K]
- T_rh = temperature of reheater metal tubes [°K]
- T_et = temperature of economizer metal tubes [°K]
- h_A = inlet air enthalpy [J/kg]

Outputs :

- Q_ir = heat transferred to the risers [J/s]
- Q_is = heat transferred by radiation to the superheater [J/s]
- Q_rs = heat transferred to the reheater [J/s]
- Q_es = heat transferred to the economizer [J/s]
- p_G = furnace air pressure [Pa]
- Q_gs = total heat transferred to the superheater [J/s]
- h_EG = enthalpy of exhaust gas from the boiler [J/kg]
- w_EG = mass flow of exhaust gas from the boiler [kg/s]
- T_g = gas temperature at the superheater [°K]
- T_gr = gas temperature at the reheater [°K]
- T_ge = gas temperature at the economizer [°K]
- T_g1 = boiler exhaust gas temperature [°K]
- y_ex = percentage excess air [%]
- x_F1 = h_EG * rho_EG [J/m^3]
- rho_EG = density of exhaust gas from the boiler [kg/m^3]

Parameters :

- K_F = chimney flow coefficient [m*s]
- K = attenuation coefficient [-]
- k_gs = experimental heat transfer coefficient to the superheater [J/(kg*°K)]
- C_gs = combustion gas specific heat capacity [(J*s)/(kg*°K)]
- k_rs = experimental heat transfer coefficient to the reheater [J/(kg*°K)]
- V_F = combustion chamber volume [m^3]
- C_F = fuel calorific value [J/kg]
- R_s = stoichiometric air/fuel ratio [-]
- gamma = content of fresh air in exhaust from gas turbine [-]
- k_es = experimental heat transfer coefficient to the economizer [J/(kg*°K)]

States :

- x_F1 = h_EG * rho_EG [J/m^3]
- rho_EG = density of exhaust gas from the boiler [kg/m^3]

Steady State Operating Condition :

Inputs :
w_F = 14.083 kg/s
w_A = 64.093 kg/s
h_G = 6.9x10^5 J/kg
w_G = 23.168 kg/s
theta = 0.88041 rad
T_st = 737.06 °K
T_rh = 743.66 °K
T_et = 412.68 °K
h_A = 2.5 x 10^5 J/kg

Parameters :
K_F = 0.001 m*s
K = 0.18
k_gs = 3532 J/(kg*°K)
C_gs = 1045 J*s/(kg*°K)
k_rs = 1.3926x10^4 J/(kg*°K)
V_F = 2000 m^3
C_F = 2.91x10^7 J/kg
R_s = 3.5
gamma= 0.1
k_es = 247.549 J/(kg*°K)

States :
x_F1 = 3.9984x10^5 J/m^3
rho_EG = 0.45052 kg/m^3

The values of outputs are:

Q_{ir} = 2.6846 x 10^7 J/s
Q_{is} = 3.6417 x 10^6 J/s
Q_{rs} = 3.1749 x 106 J/s
Q_{es} = 1.2465 x 106 J/s
T_{ge} = 727.94 °K
T_{g1} = 716.17 °K
X_{F1} = 3.9984 x105 J/m^3
p_G = 1.013 x 10^5 Pa

ρ_{EG} = 0.45052 kg/m^3
h_{EG} = 8.8752 x 10^5 J/kg
W_{EG} = 101.3 kg/s
T_g = 781.75 °K
T_{gr} = 757.94 °K
y_{ex} = 34.935 %

A1.1.2 The Drum Part

Inputs :

- h_e = specified enthalpy of water from the economizer [J/kg]
- v_dow = volumetric water flow to the downcomer [m^3/s]
- w_e = water flow from the economizer [kg/s]
- Q_ir = heat flow from the furnace [J/s]
- w_v = steam flow to the superheater [kg/s]

Outputs :

- p_v = drum outlet steam pressure [Pa]
- rho_v = drum outlet steam density [kg/m^3]
- h_v = drum outlet steam specific enthalpy [J/kg]
- h_r = liquid-vapour mixture specific enthalpy [J/kg]
- T_rt = risers metal tube temperature [$^{\circ}$K]
- w_r = liquid vapour mixture mass flow from the risers [kg/s]
- rho_w = drum water density [kg/m^3]
- w_d = water mass flow to the downcomer [kg/s]
- m_ dl = drum liquid mass [kg]
- L = drum water level [m]
- x_qa = steam quality [-]
- T_w = drum water temperature [$^{\circ}$K]

Parameters :

- V = volume of the drum [m^3]
- k_ec = evaporation coefficient [kg/($^{\circ}$K*s)]
- r = drum radius [m]
- w_ec0 = steady-state evaporation constant [kg/s]
- k_r = experimental heat transfer coefficient [J/(s*°K^3)]
- V_r = risers volume [m^3]
- M_r = mass of risers metal tubes [kg]
- C_rt = metal specific heat [J/(kg*$^{\circ}$K)]
- A_r = not used in version 1
- L_r = not used in version 1
- f_r = not used in version 1
- D_r = not used in version 1
- tau_r = mass flow time constant [s]

States :

- m_dl = drum liquid mass [kg]
- x_D1 = h_w * m_dl
- x_D2 = rho_v * v_v where V_v = volume of steam in the drum
- h_r = liquid-vapour mixture specific enthalpy [J/kg]
- T_rt = risers metal tube temperature [$^{\circ}$K]
- w_r = liquid-vapour mixture mass flow from the risers [kg/s]

Steady State Operating Condition :

Inputs :

h_e = 5.6217 J/kg
v_dow = 0.71556 m^3/s
w_e = 12 kg/s
Q_ir = 2.6546 x 10^7 J/s
w_v = 12 kg/s

C_rt = 481.4 J/(kg*°K)
A_r = 0.893 m^2
L_r = 7.315 m
f_r = 0.044
D_r = 1.066 m
tau_r = 1 s

Parameters :

V = 9.253 m^3
k_ec = 0.6124 kg/(°K*s)
r = 0.61 m
w_ec0 = 0 kg/s
k_r = 444.2 J/(s*°K^3)
V_r = 6.53 m^3
M_r = 2.25x10^4 kg

States :

m_dl = 3817.6 kg
x_D1 = 4.2708 x 10^9 J
x_D2 = 100.396 kg
h_r = 1.17 x 10^6 J/kg
T_rt = 567.9 °K
w_r = 564.11 kg/s

The values of outputs are:

p_v = 4.5417 x 10^6 Pa
ρ_v = 22.763 kg/m^3
h_v = 1.1663 x 10^6 J/kg
T_{rt} = 567.9 °K
w_r = 564.11 kg/s
ρ_w = 788.34 kg/m^3

W_d = 564.11 kg/s
m_{dl} = 3817.6 kg
L = 4.1425 m
x_{qa} = 0.02334
T_w = 526.76 °K

A1.1.3 Superheater and Attemporator

Inputs :

- W_a = attemporation water flow [kg/s]
- w_s = steam flow from the superheater [kg/s]
- p_v = steam drum pressure [Pa]
- rho_v = density of saturated steam from the drum [kg/m^3]
- Q_gs = heat flow from the furnace [J/s]
- h_v = specific enthalpy of saturated steam from the drum [J/kg]
- h_a = specific enthalpy of attemporation water [J/kg]

Outputs :

- w_v = drum outlet steam pressure [Pa]
- T_st = superheater metal tube temperature [°K]
- p_s = pressure of superheated steam [Pa]
- T_s = temperature of superheated system [°K]
- h_f = specific enthalpy of evaporation [J/kg]
- Q_s = heat transferred to the steam [J/s]

- rho_s = density of superheated steam [kg/m^3]
- x_s1 = h_s * rho_s
- h_s = specific enthalpy of superheated steam [J/m^3]

Parameters :
- f_s = superheater friction coefficient [m^{-4}]
- k_s = experimental heat transfer coefficient [J/(kg*°K)]
- V_s = superheater volume [m^3]
- M_s = superheater mass [kg]
- C_st = heat capacitance of superheater tubes [J/(kg*°K)]
- Cp_ref = ideal gas reference specific heat [J/(kg*°K)]
- T_ref = ideal gas reference temperature [°K]
- h_ref = ideal gas reference specific enthalpy [J/kg]

States :
- rho_s = density of superheated steam [kg/m^3]
- T_st = superheater metal tube temperature [°K]
- x_s1 = h_s * rho_s

Steady State Operating Conditions :

Inputs :
w_a = 0 kg/s
w_s = 12 kg/s
p_v = 4.5417 x 10^6 Pa
rho_v = 22.763 kg/m^3
Q_gs = 6.1626 x 10^6 J/s
h_v = 2.7977 x 10^6 J/kg
h_a = 5.5217 x 10^5 J/kg

Parameters :
f_s = 2615 m^{-4}
k_s = 4.37 x 10^4 J/(kg °K)
V_s = 8.462 m^3
M_s = 1.04 x 10^4 kg
C_st = 481.4 J/(kg °K)
Cp_ref = 2330 J/(kg °K)
T_ref = 723.15 °K
h_ref = 3.32 x 10^6 J/kg

States :
rho_s = 13.662 kg/m^3
T_st = 737.66 °K
x_s1 = 4.5244 x 10^7 J/m^3

The values of outputs are:

w_v = 12 kg/s
T_{st} = 737.06 °K
p_s = 45.251 x 10^5 Pa
T_s = 717.72 °K
h_f = 1.8428 x 10^6 J/kg
Q_s = 6.1714 x 10^6 J/s
ρ_s = 13.662 kg/m^3
x_{s1} = 4.5244 x 10^7 J/m^3
h_s = 3.3117 x 10^6 J/kg

A1.1.4 The Reheater

Inputs :
- p_ri = pressure of steam at the inlet to the reheater [Pa]
- w_ri = flow of steam at the inlet to the reheater [kg/s]
- T_ri = inlet steam temperature [°K]
- Q_rs = heat flow from the furnace [J/s]
- h_ri = specific enthalpy of inlet steam [J/kg]

Outputs :
- T_rh = reheater metal tube temperature [°K]
- p_ro = outlet steam pressure [Pa]
- T_r = reheater steam temperature [°K]
- h_ro = specific enthalpy of outlet steam [J/kg]
- Q_rh = heat transferred to the steam [J/s]
- rho_rh = steam density in the reheater [kg/m^3]
- x_RH1 = h_ro * rho_rh
- w_ro = flow of steam at the outlet from the reheater [kg/s]

- **Parameters :**
- k_rh = experimental heat transfer coefficient [J/(kg* °K)]
- V_rh = reheater volume [m^3]
- M_r = reheater mass [kg]
- C_rh = heat capacitance of superheater tubes [J/(kg* °K)]
- Cp_ref = ideal gas reference specific heat [J/(kg* °K)]
- T_ref = ideal gas reference temperature [°K]
- h_ref = ideal gas reference specific enthalpy [J/kg]

States :
- rho_rh = steam density in the reheater [kg/m^3]
- T_rh = reheater metal tube temperature [°K]
- x_RH1 = h_ro * rho_rh
- w_ro = outlet steam mass flow [kg/s]

Steady State Operating Conditions :

Inputs :

p_ri = 1.3896 x 10^6 Pa
w_ri = 10.459 kg/s
T_ri = 601.69 °K
Q_rs = 3.1748 x 10^6 J/s
h_ri = 3.0298 x 10^6 J/kg

Parameters :

f_r = not used

k_rh = 2.95 x 10^4 J/(kg °K)
V_rh = 10 m^3
M_r = 7000 kg
C_rh = 481 J/(kg °K)
Cp_ref = 2200 J/(kg °K)
T_ref = 723.16 °K
h_ref = 3.3244 x 10^6 J/kg

States :

rho_rh = 3.8835 kg/m^3

$T\_rh = 743.66$ °K
$x\_RH1 = 1.2945 \times 10^7$ J/m^3
$w\_ro = 10.459$ kg/s

The values of steady state outputs are:

$T\_rh = 743.66$ °K
$p\_ro = 1.3034 \times 10^6$ Pa
$T\_r = 727.25$ °K
$h\_ro = 3.333 \times 10^6$ J/kg
$Q\_rh = 3.1744 \times 10^6$ J/s
$\rho\_rh = 3.8835$ kg/m^3
$x\_RH1 = 1.2945 \times 10^7$ J/m^3
$w\_ro = 10.459$ kg/s

A1.1.5 Boiler Control

Inputs :

- p_des = output power set point [Pa]
- p_s = superheated steam pressure [W]
- L_sp = drum water level set point[m]
- L = drum water level [m]
- p_g_sp = furnace air pressure set point [Pa]
- p_g = furnace air pressure [Pa]
- T_s_sp = superheated steam temperature set point [°K]
- T_s = superheated steam temperature [°K]
- T_rh_sp = reheated steam temperature set point [°K]
- T_rh = reheated steam temperature [°K]

Outputs :

- w_f = fuel flow to the furnace [kg/s]
- v_{pos_e} = valve position in the feedwater system (economizer) [-]
- w_A = air flow to the furnace [kg/s]
- v_{pos_att} = valve position in the feedwater system (attemporator) [-]
- theta = tilt angle [coefficient]

Steady State Operating Condition :

Inputs :

$P_s = P_{ssp} = 45.251 \times 10^5$ Pa
$L = L_{sp} = 4.1425$ m
$pg = p_{gsp} = 1.013 \times 10^5$ Pa
$Ts = T_{ssp} = 717.72$ °K
$Trh = T_{rhsp} = 743.66$ °K

The values of steady state outputs are:

$W_f = 14.083$ kg/s
$v_{pos\_e} = 0.55056$
$W_A = 64.093$ kg/s
$v_{pos\_att} = 0$
theta = 0.88041

A1.2 Gas Turbine

A1.2.1 Fuel System

Input :
- Per unit fuel demand

Output :
- Fuel flow [kg/s]

Parameters :
- τ_L = Transport delay [s]
- T_{vp} = Valve positioner time constant [s]
- T_{fs} = fuel system volumetric lag [s]

Steady state operating condition:
Input :
Per unit fuel demand = 0.61648

The value of steady state output is as follows:
Wf = 1.8494 kg/s

A1.2.2 Compressor

Inputs :
- w_a = inlet air flow [kg/s]
- rho_i = inlet air density [kg/m^3]
- p_cin = inlet air pressure [Pa]
- T_cin = inlet air temperature [°K]

Outputs :
- p_cout = outlet air pressure [Pa]
- T_cout = outlet air temperature [°K]
- P_c = power consumption [W]

Parameters :
- gamma_a = ratio of specific heats for air
- eta_inf_c = polytropic efficiency of compressor
- A_0 = exit flow area [m^2]
- eta_trans = transmission efficiency
- R_air = ideal gas constant for air [J/(kg °K)]
- Cp_air = specific heat at constant pressure for air [J/(kg °K)]

- r_c_old = nominal outlet/.inlet pressure ratio

Steady state operating conditions :

Inputs :

$W_a = 46.137$ kg/s

$\rho_i = 1.21$ kg/m^3

p_cin = 1×10^5 Pa

T_cin = 288 °K

Parameters :

$\gamma_a = 1.4$

$\eta_{Trans} = 0.99$

$\eta_\infty = 0.9$

R_air = 287 J/(kg °K)

$A_\phi = 0.01$ m^2

Cp_air = 1005 J/(kg °K)

$r_{cold} = 10$

The values of steady state outputs are:

$p_{cout} = 1.0033 \times 10^6$ Pa

$T_{cout} = 598.8$ °K

$P_c = 1.4549 \times 10^7$ W

A1.2.3 Combustion Chamber

Inputs :

- p_cout = compressor outlet air pressure [Pa]
- T_cout = compressor outlet air temperature [°K]
- w_a = air flow to combustor from compressor [kg/s]
- w_f = fuel flow to combustor [kg/s]
- w_is = injected steam flow to combustion [kg/s]
- T_is = temperature of injected steam [°K]

Outputs :

- w_g = outlet gas flow from combustor [kg/s]
- T_Tin = outlet gas pressure [°K]
- p_Tin = outlet gas pressure [Pa]
- gc_nox = Nox content in outlet gas [ppm]
- gc_co = Co content in outlet gas [ppm]
- h_Tin = outlet gas enthalpy [J/kg]

Parameters :

- cp_g = specific heats at constant pressure for outlet gas [J/(kg °K)]
- cp_a = specific heat at constant pressure for air [J/(kg °K)]

- cp_s = specific heat at constant pressure for steam [J/(kg °K)]
- deltah_25 = fuel enthalpy of combustion at reference temperature [J/kg]
- k1 = combustion chamber pressure drop empirical coefficient
- k2 = combustion chamber pressure drop empirical coefficient
- R_cg = ideal gas constant for combustion gases [J/(kg °K)]
- A_m = combustion chamber cross-section area [m^2]
- h_ref = reference combustion gas enthalpy [J/kg]
- T_ref = reference combustion gas temperature [°K]

Steady state operating conditions:

Inputs :

$p_{cout} = 1.0033 \times 10^6$ Pa

$T_{cout} = 598.8$ °K

$T_{is} = 601.69$ °K

$W_a = 46.137$ kg/s

$W_f = 1.8484$ kg/s

$W_{is} = 0.18566$ kg/s

Parameters :

$C_{pg} = 1144$ J/(kg °K)

k1 = 1

$A_m = 1$ m^2

$\Delta H_{25} = -4.0 \times 10^7$ J/kg

$R_{cg} = 287$ J/(kg °K)

$H_{ref} = 1.2041 \times 10^6$ J/kg

$T_{ref} = 1.000$ °K

The values of steady state outputs are:

$w_g = 46.939$ kg/s

$T_{Tin} = 1892$ °K

$P_{Tin} = 1.0027 \times 10^6$ Pa

$g_{cnox} = 90.76$ ppm

$g_{cco} = 5.826$ ppm

$h_{Tin} = 2.2267 \times 10^6$ J/kg

A1.2.4 The Turbine

Inputs :

- w_g = inlet gas flow from combustor [kg/s]
- T_tin = inlet gas temperature [°K]
- p_tin = inlet gas pressure [Pa]
- p_c = compressor power consumption [W]
- h_tin = inlet gas enthalpy [J/kg]

Outputs :

- P_mech = net mechanical power produced [W]
- T_tout = exhaust gas temperture [°K]
- p_tout = exhaust gas pressure [Pa]
- h_tout = exhaust gas enthalpy [J/kg]

Parameters :

- eta_inf_t = turbine polytropic efficiency
- gamma_cg = specific heat ratio for combustion gases
- A_t0 = turbine exit flow area [m^2]
- R_cg = ideal gas constant for combustion gases [J/(kg °K)]
- c_pcg = specific heat at constant pressure for combustion gases [J/(kg °K)]
- R_t_old = nominal outlet/inlet presure ratio for turbine.

Steady state operating condition:

Inputs :

$W_g = 46.939$ kg/s

$T_{tin} = 1892$ °K

$p_{tin} = 1.0027 \times 10^6$ Pa

$P_c = 1.4547 \times 10^7$ W

$h_{tin} = 1.1804 \times 10^6$ J/kg

$\eta_{\infty t} = 0.9$

$A_{t\phi} = 0.14$ m^2

$C_{pg} = 1144$ J/(kg °K)

$\gamma_{cg} = 1.333$

$R_{cg} = 287$ J/(kg °K)

$r_{told} = 0.1$

Parameters :

The values of steady state outputs are:

$P_{mech} = 3.3896 \times 10^7$ W

$T_{tout} = 1015.8$ °K

$P_{tout} = 6.312 \times 10^4$ Pa

$h_{tout} = 1.2211 \times 10^6$ J/kg

A1.2.5 The Gas Turbine Controller

Inputs :

- T_exhaust_sp = exhuast temperature set point [°K]
- ω_sp = shaft speed set point [rad/s]
- $p_{mech-sp}$ = mechanical power set point [W]
- min_fuel_sp = minimum fuel set point [kg/s]
- $g_{cnox-sp}$ = Nox content in exhaust set point [ppm]
- T_exhaust = exhaust temperature [°K]
- w_pu = per unit shaft speed
- $p_{mech-pu}$ = per unit mechanical power
- g_{cnox} = exhaust gas nox content [ppm]

Outputs :

- W_air_dem = demanded air flow [kg/s]
- F_dem_pu = demanded fuel flow per unit
- W_{is} = injected steam flow [kg/s]

Parameters :

- radiation shield gain and time constant [s]
- thermocouple time constant [s]
- Nox analyser time constant [s]
- PID controller gains, initial conditions
- controller deadbands and limiter settings.

Steady state operating conditions :

Inputs :

$T_{ex\text{-}sp} = T_{ex} = 1015.8$ °K

$\omega_{\text{-}sp} = 157.08$ rad/s

$\omega\text{-pu} = 1$

$p_{mech\text{-}sp} = 3.3896 \times 10^7$ W

$p_{mech\text{-}pu} = 1$

$g_{cnox\text{-}sp} = g_{cnox} = 90.76$ ppm

The values of steady state outputs are:

$W_{airdem} = 46.137$ kg/s

$F_{dem\text{-}pu} = 0.61648$

$W_{is} = 0.18566$ kg/s

A1.3 The Steam Turbine

A1.3.1 The High Pressure Section

Inputs :

- w_in = inlet steam flow from boiler superheat section [kg/s]
- p_in = inlet steam pressure [Pa]
- T_in = inlet steam flow temperature [°K]
- h_in = inlet steam flow enthalpy [J/kg]
- rho_in = inlet steam flow density [kg/m^3]

Outputs :

- w_ouhp = outlet steam flow [kg/s]
- p_ouhp = outlet steam pressure [Pa]
- h_ouhp = outlet steam enthalpy [J/kg]
- P_hp = mechanical power developed in HP section [W]
- T_ouhp = outlet steam temperature [°K]

Parameters :
- eta_inf_hp = HP section polytropic efficiency
- A_hp= HP section exit flow area [m^2]
- V_hp = HP section steam storage volume [m^3]
- f_hp = REDUNDANT in Version 1
- cp_hp = specific heat at constant pressure for high pressure steam [J/(kg °K)]
- tau_hp = HP section mass flow time constant [s]
- r_t_old = nominal outlet/inlet pressure ratio for HP section

States :
- rho_o = steam density in HP section steam storage [kg/m^3]
- x_o = rho_o*h_o steam energy per unit volume in HP section steam storage [J/m^3]
- w_ouhp = outlet steam flow [kg/s]

Steady state operating condition:

Inputs:

W_{in} = 12.0 kg/s

ρ_{in} = 13.662 kg/m^3

p_{in} = 4.5251x10^6 Pa

h_{in} = 3.3117x10^6 J/kg

T_{in} = 717.72 °K

Parameters :

$\eta_{\infty hp}$ = 0.8

A_{hp} = 0.003 m^2

f_{hp} = NOT USED

C_{php} = 2430 J/(kg °K)

V_{hp} = 5.664 m^3

τ_{hp} = 1 s

r_{told} = 0.333

States :

ρ_o = 13.353 kg/ m^3

x_o = 4.421 x10^7 J/ m^3

W_{ouhp} = 12 kg/s

The steady state outputs are as follows:

W_{ouhp} = 12 kg/s

P_{ouhp} = 1.389 x10^6 Pa

h_{ouhp} = 3.0298 x10^6 J/kg

P_{hp} = 3.3746 x10^6 W

T_{ouhp} = 601.69 °K

A1.3.2 The Intermediate Pressure Section

Inputs :
- w_inip = inlet steam flow from boiler reheater section [kg/s]
- p_o = inlet steam pressure [Pa]
- T_o = inlet steam temperature [°K]

- h_o = inlet steam enthalpy [J/kg]
- rho_o = inlet steam density [kg/m^3]

Outputs :
- p_ouip = outlet steam pressure [Pa]
- h_ouip = outlet steam enthalpy [J/kg]
- P_ip = mechanical power developed in IP section [W]
- T_ouip = outlet steam temperature [°K]
- rho_ouip = outlet steam density [kg/m^3]
- w_ouip = outlet steam flow [kg/s]

Parameters :
- eta_inf_ip = IP section polytropic efficiency
- A_ip = IP section exit flow area [m^2]
- cp_ip = specific heat at constant pressure for intermediate pressure steam [J/(kg °K)]
- r_t_old = nominal outlet/inlet pressure ratio for IP section

Steady state operating condition:

Inputs:

$W_{inip} = 10.459$ kg

$h_o = 3.3334 \times 10^6$ J/kg

$p_o = 1.3034 \times 10^6$ Pa

$\rho_o = 3.8835$ kg/m^3

$T_o = 727.25$ °K

$r_{told} = 0.358$

Parameters :

$\eta_{\infty ip} = 0.85$

$C_{pip} = 2230$ J/(kg °K)

$A_{ip} = 0.009$ m^2

The values of steady state outputs are:

$p_{ouip} = 4.777 \times 10^5$ Pa

$h_{ouip} = 3.0709 \times 10^6$ J/kg

$P_{ip} = 2.7453 \times 10^6$ W

$T_{ouip} = 609.54$ °K

$\rho_{ouip} = 1.6982$ kg/m^3

$W_{ouip} = 10.459$ kg/s

A1.3.3 The Low Pressure Section

Inputs :
- w_inlp = inlet steam flow from IP section [kg/s]
- p_inlp = inlet steam pressure [Pa]
- T_inlp = inlet steam temperature [°K]
- h_inlp = inlet steam enthalpy [J/kg]
- rho_inlp = inlet steam density [kg/m^3]

Outputs :

- w_oulp = outlet steam flow [kg/s]
- p_oulp = outlet steam pressure [Pa]
- h_oulp = outlet steam enthalpy [J/kg]
- P_lp = mechanical power developed in LP section [W]
- T_oulp = outlet steam pressure [°K]

Parameters :

- eta_ing_lp = LP section polytropic efficiency
- A_lp = LP section exit flow area [m^2]
- V_lp = LP section steam storage volume [m^3]
- cp_lp = specific heat at constant pressure for low pressure steam [J/(kg °K)]
- tau_lp = LP section mass flow tine constant [s]
- r_t_old = nominal outlet/inlet pressure ratio for LP section

States :

- rho_o = steam density in LP section steam storage [kg/m^3]
- x_o = rho_o*h_o steam energy per unit volume in LP section
- p_o = steam storage [J/m^3]
- w_oulp = outlet steam flow [kg/s]

Steady state operating condition:

Inputs:

w_{inlp} = 10.459 kg/s

$h_{inlp} = 3.0709 \times 10^6$ J/kg

$p_{inlp} = 4.777 \times 10^5$ Pa

ρ_{inlp} = 1.5982 kg/m^3

T_{inlp} = 609.54 °K

Parameters :

$\eta_{\infty lp}$ = 0.85

V_{lp} = 5.664 m^3

A_{lp} = 0.05 m^2

C_{plp} = 2005 J/(kg °K)

τ_{lp} = 1 s

r_{told} = 0.0286

States :

w_{oulp} = 10.459 kg/s

$p_{oulp} = 3.7089 \times 10^4$ Pa

$h_{oulp} = 2.5921 \times 10^6$ J/kg

$P_{lp} = 5.0076 \times 10^6$ W

T_{oulp} = 375.98 °K

A1.3.4 The Steam Turbine Controller

Inputs :

- ω_{pu} = per unit shaft speed [rad/s]

- ω_{opu} = rated speed [rad/s]
- p_{mechpu} = per unit mechanical power
- p_{inhp} = superheated steam inlet pressure [Pa]

Outputs :
- $x_{dem\text{-}cv}$ = demanded governor valve steam position
- $x_{dem\text{-}by}$ = demanded bypass valve steam position
- $x_{dem\text{-}iv}$ = demanded intercept valve steam position

Steady state operating condition:
Inputs:
$\omega_{pu} = 1$
$P_{inhp} = 4.5251 \times 10^6$ Pa
$\omega_{opu} = 1$
$P_{mechpu} = 1$

The values of steady state outputs are:
$x_{dem\text{-}cv} = 0.95268$
$x_{dem\text{-}by} = 0.0$
$x_{dem\text{-}iv} = 0.0$

A1.4 The Condenser

A1.4.1 The Shell

Inputs :
- T_st = steam temperature [°K]

Outputs :
- Q_sh = heat flow to shell from steam [J/s]
- T_sh = shell metal temperature [°K]

Parameters :
- H_sh = heat transfer coefficient for shell metal [J/(s* °K *m^2)]
- A_sh= surface area of the shell [m^2]
- m_sh = mass of the shell [kg]
- C_psh = specific heat of shell material [J/(kg* °K)]

States :
- T_sh = steam metal temperature [°K]

Steady state operating condition:

Inputs:

T_st = 308.99 ^{o}K

m_sh = 480 kg

C_{-psh} = 3200 J/(kg ^{o}K)

Parameters :

H_sh = 1140 J/(s ^{o}K m^2)

A_sh = 49.8 m^2

States :

Y_sh = 308.99 ^{o}K

The values of outputs are

Q_{sh} = 6.364 J/s

T_{sh} = 308.9 ^{o}K

A1.4.2 The Tube Material

Inputs :

- T_st = steam temperature [°K]
- T_fi(i) = average liquid temperature in the i-th section (i-1, ,N) [°K]

Outputs :

- Q_hfi(i) = heat flow from tubes to cooling liquid in ith section (i-1,..,N) [J/s]
- T_t = temperature of tubes [°K]
- fi_t = overall heat flow coefficient for the tubes [(°K *s)/J]
- fi_ti = heat flow coefficient for a section [(°K *s)/J]
- Q_{st} = heat flow from steam to tubes [J/s]

Parameters :

- Nin_t = number of tubes [-]
- N = number of sections in liquid equation [-]
- r_o = outside radius of tubes [m]
- r_i = inside radius of tubes [m]
- k_t = thermoconductivity of tube walls [J/(°K *s*m)]
- L_t = length of tubes [m]
- A_ot = surface area of tubes [m^2]
- H_st = heat transfer coefficient for tubes material steam to tubes [J/(°K *s*m^2)]
- H_tl = heat transfer coefficient for tubes material steam to tubes [J/(°K *s*m^2)]
- m_t = mass of one tube [kg]
- c_pt = specific heat of tubes material [J/(kg* °K)]

States :

- T_t = temperature of tubes [°K]

Steady state operating condition:

Inputs:

$T\_st = 308.99$ °K

$T\_fi(i) = 295.52$ °K

$T\_fi(2) = 299.31$ °K

$T\_fi(N) = 301.1$ °K

Parameters :

$N\_t = 80$

$N = 3$

$r_o = 0.08$m

$r_i = 0.078$m

$k\_t = 2.2$ J/(s °K m)

$L\_t = 3.7$ m

$A\_ot = 1.85$ m^2

$H\_st = 28$ J/(°K s m^2)

$H\_tl = 28$ J/ (°K s m^2)

$m\_t = 200$ kg

$c\_pt = 400$ J/(kg °K)

States :

$T\_t = 302.75$ °K

The values of steady state outputs are:

$Q_{hfi(1)} = 1.4875 \times 10^7$ J/s

$Tt = 302.75$ °K

$Q_{st} = 2.5534 \times 10^7$ J/s

$Q_{hfi(2)} = 7.1908 \times 10^6$ J/s

$f_{i-t} = 2.4441 \times 10^{-4}$ °Ks/J

$Q_{st} = 2.5534 \times 10^7$ J/s

$Q_{hfi(n)} = 3.4674 \times 10^6$ J/s

$f_{i-ti} = 7.3581 \times 10^{-4}$ °Ks/J

A1.4.3 The Steam Equations

Inputs :

- w_st = inlet flow of steam [kg/s]
- h_st = specific enthalpy of saturated steam entering the condenser [J/kg]
- Q_sh = heat flow from steam to the shell [J/s]
- Q_st = heat flow from steam to the tubes [J/s]

Outputs :

- T_st = steam temperature[°K]
- p_cond = pressure of condensate at saturation conditions [Pa]
- W_cond = condensate mass flow [kg/s]
- h_cond = condensate enthalpy [J/kg]

Parameters :

- V = volume of steam in the shell [m^3]

States :

- T_st = steam temperature [°K]

Steady state operating condition:

Inputs:

w_st = 10.459 kg/s

h_st = 2.5921×10^6 J/kg

W_sh = 6.364 J/s

Q_st = 2.5534×10^7 J/s

Parameters :

V= 48.8 m^3

States :

T_st = 308.99 °K

The values of outputs are as follows :

T_st = 308.99 °K

$P_{cond} = 5.8667 \times 10^3$ Pa

$W_{cond} = 10.459$ kg/s

$h_{cond} = 1.5071 \times 10^5$ J/kg

A1.4.4 The Liquid

Inputs :

- w_f = cooling liquid mass flow [kg/s]
- T_fo = cooling liquid inlet temperature [°K]
- Q_hfi(i) = heat flow from tubes to cooling liquid at i-th section (i=1,.., N) [J/s].

Outputs :

- deltap_f = overall liquid pressure drop [Pa]
- T_hfi(i) = cooling liquid outlet temperature (i=1, ,N) [°K]
- T_fi_dash(i) = outlet liquid temperature from i-th section (i=0,...,N-1) [°K]

Parameters :

- N = number of sections
- rho_f = density of cooling liquid [kg/m^3]
- c_pf = specific heat at constant pressure [J/(kg*°K)]
- V_fi = volume of liquid in j-th section [m^3]
- f_f = friction coefficient [m^{-4}]

States :

- T_fi(i+1) = outlet liquid temperature from i-th section (i=0,...,N-1) [°K]

Steady state operating condition:

Inputs:

W_f = 464.32 kg/s

T_{fo} = 288 °K

Q_hfi(i) = 1.4875×10^7 J/s

Q_hfi(2) = 7.1908×10^6 J/s

Q_hfi(N) = 3.4564×10^6 J/s

Parameters :

N = 100

rho_f = 4200 kg/m^3

c_pf - 127 J/(kg °K) f_f = 2615 m^{-4}

V_fi = 3 m^3

States :

T_fi(i+1) = [295.616 °K , 299.309 °K, 301.096 °K]

The values of steady state outputs are:

Δp_f = 1.6829x10^5 Pa

$T_{fi\text{-}dash(1)}$ = 291.81 °K

$T_{fi\text{-}dash(2)}$ = 297.46 °K

$T_{fi\text{-}dash(N)}$ = 200.46 °K

$T_{fi\ (1)}$ = 295.62 °K

$T_{fi\ (2)}$ = 299.31 °K

$T_{fi\ (N)}$ = 301.1 °K

A1.5 Feedwater System

A1.5.1 Economizer

Inputs :

- p_ei = inlet water pressure [Pa]
- w_ei = inlet water mass flow [kg/s]
- w_eo = outlet water mass flow [kg/s]
- Q_es = heat flow from the furnace [J/s]

Outputs :

- T_et = temperature of economizer tubes [°K]
- p_eo = economizer outlet water pressure [Pa]
- T_eo = economizer outlet water temperature [°K]
- rho_e = density of liquid in the economizer [kg/m^3]
- Q_e = heat flow to the economiser liquid [J/kg]
- X_e1 = rho_e * he_o
- h_eo = outlet water specific enthalpy [J/kg]
- h_ei = inlet water specific enthalpy [J/kg]

Parameters :

- f_e = economizer tubes friction coefficient - not used
- k_e = heat transfer coefficient [J/(kg*°K)]
- V_e = total volume of economizer [m^3]
- M_e = mass of economizer tubes [kg]
- C_e = heat capacitance of economzier tubes [J/(kg*°K)]

States :

- rho_e = density of liquid in the economiser [kg/m^3]

- T_et = temperature of economizer tubes [°K]
- $X\_e1 = rho\_e * he\_o$

Steady state operating condition:

Inputs:

p_ei = 1.8339×10^5 Pa

w_ei = 12.5 kg/s

w_eo = 12 kg/s

Q_es = 1.2465×10^5 J/s

Parameters :

f_e = NOT USED

k_e = 43700 J/(kg °K)

V_e = 3 m^2

C_e = 7000 kg

Outputs :

T_t = 412.68 °K

p_eo = 2.97891×10^5 Pa

T_eo = 408.91 °K

rho_e = 1188.7 kg/m^3

Q_e = 1.2429×10^6 J/s

X_e1 = 6.6826×10^8 J/m^3

h_eo = 5.6217×10^5 J/kg

h_ei = 4.6392×10^5 J/kg

States :

rho_e =1188.7 kg/m^3

T_et = 412.68 °K

X_e1 = 6.6826×10^8 J/m^3

A1.5.2 The Deaerator

Inputs :

- w_f = mass flow of outlet feedwater [kg/s]
- w_c = mass flow of condensate (inlet) [kg/s]
- w_m = mass flow of makeup water [kg/s]
- w_x = mass flow of steam extracted from the turbine [kg/s]
- h_c = specific enthalpy of condensate [J/kg]
- h_m = specific enthalpy of makeup water [J/kg]
- h_x = specific enthalpy of steam [J/kg]

Outputs :

- p_fo = outlet feedwater pressure [Pa]
- l = deaerator liquid level [m]
- h_f = specific enthalpy of outlet feedwater [J/kg]
- p_f = pressure of water in the two-fase region [Pa]
- rho_f = density of feedwater in the deaerator [kg/m^3]
- V_f = temperature of liquid inside the deaerator [°K]
- m_f = mass of liquid in the deaerator [kg]
- X_01 = h_f * m_f

Parameters :

- V_dae = deaerator internal volume [m^3]
- r = deaerator radius [m]

States :

- m_f = mass of liquid in the deaerator [kg]
- X_01 = h_f * m_f

Steady state operating condition:

Inputs :

w_f = 12.5 kg/s

w_c = 10.459 kg/s

w_m = 0.67732 kg/s

w_x = 1.3633 kg/s

h_c = 1.5071 x 10^5 J/kg

h_m = 8.4 x 10^4 J/kg

h_x = 3.0298 x 10^6 J/kg

Parameters :

V_dae = 3 m^3

r = 0.5 m

Outputs:

p_fo = 5.397 x 10^4 Pa

l = 1.913 m

h_f = 3.35 x 10^5 J/kg

p_f = 3.54759 x 10^4 Pa

rho_f = 970.42 kg/m^3

T_f = 354.72 kg/m^3

V_f = 1.5025 m^3

m_f = 1458 kg

X_01 = 4.8843 x 10^8 J/m^3

States :

m_f = 1458kg

X_01 = 4.8843 x 10^8 J/m^3

A1.5.3 The Pump

Inputs :

- N_dem = pump speed demand [rad/s]
- p_in = inlet pressure [Pa]
- rho_in = inlet density [kg/m^3]

Outputs :

- w_out = outlet mass flow [kg/s]
- P_w = pump power consumption [J/s]
- p_out = outlet pressure [Pa]
- rho_out = outlet density [kg/m^3]

Parameters :

- k_1 = a pump construction constant [m^2]
- k_2 = a pump construction constant [m^{-4}]
- k_3 = pump load coefficient [m^{-4}]
- eta = pump efficiency [%]
- rho_av = average fluid density [kg/m^3]
- k_2 = a pump construction constant [m^{-4}]
- delta_ps = static pressure rise across the pump [Pa]

Steady state operating condition:

Inputs:

N = 211.92 rad/s

p_in = 6.397×10^4 Pa

rho_in = 970.42 kg/m^3

rho_out = 970 kg/m^3

Outputs :

w_out = 12.3 kg/s

p_w = 1.71×10^7 W

p_out = 1.8339×10^5 Pa

Parameters :

k_1 = 0.003 m^3

k_2 = 0

k_3 = 2.39×10^5 m^{-4}

eta = 0.9

rho_av =974 kg/m^3

delta_ps = 8.08×10^5 Pa

A1.5.4 The Valve

Input :

- x_pos = valve stem position (0 : 1)
- rho_1 = liquid density [kg/m^3]
- p_ i = inlet liquid pressure [Pa]

Outputs :

- w_1 = liquid mass flow [kg/s]
- p_o = outlet liquid pressure [Pa]

Parameters :

- c_v1 = constant valve construction coefficient [m^2]
- delta_p = pressure drop across the valve [Pa]

Steady state operating condition:

Inputs:

x_pos = 0.55059

rho_1= 1188.7 kg/m^3

p_i = 2.9781x10 Pa

Outputs :

w_1= 12 kg/s

p_o = 2.8781×10^5 Pa

Parameters :

c_v1 = 0.005 m^2

delta_p = 1×10^4 Pa

A1.5.5 Feedwater System Controller

Inputs :
- l_sp = deaerator water level set point [m]
- l = deaerator water level [m]
- p_f_sp = deaerator water pressure set point [Pa]
- p_f_ = deaerator water pressure [Pa]

Outputs :
- w_m = mass flow of makeup water [kg/s]
- w_x = mass flow of steam (extracted from the turbine) [kg/s]

Steady state operating condition:

Inputs :

$L = L_{sp}$ = 1.913 m

$p_f = p_{fsp}$ = 4.5759 $\times 10^5$ Pa

Outputs :

w_m = 0.67861 kg/s

w_x = 1.3633 kg/s

A1.6 Electrical Generator

Inputs :
- Pmech_pu = per unit mechanical shaft power
- Pelec_pu = per unit electrical load
- ω_ϕ = per unit reference rotor speed

Output :
- ω_{pu} = per unit rotor speed

Parameters :
- M_{eq} = equivalent rated angular momentum of machine
- D_{eq} = equivalent electrical damping coefficient

Steady state operating condition:

Inputs :

Pmech_pu = 1

Pelec_pu = 1

ω_ϕ = 157.08 rad/s

Parameters :
$M_{eq} = 4.5$
$D_{eq} = 0.8$

Output :
$\omega_{pu} = 1.0$

A1.7 Gas Merge and Splitting Process

Inputs :
- F-i1 = inlet gas flow 1 [kg/s]
- F-i2 = inlet gas flow 2 [kg/s]
- F-b1set = gas flow to boiler 1 ; set point
- F-b2set = gas flow to boiler 2 ; set point
- h-i1 = inlet gas flow 1 enthalpy [J/kg]
- h-i2 = inlet gas flow 2 enthalpy [J/kg]

Outputs :
- F_{b1} = gas flow to boiler 1 [kg/s]
- F_{b2} = gas flow to boiler 2 [kg/s]
- h = exit flows enthalpy [J/kg]
- F_f = flue mass flow [kg/s]

Steady state operating conditions :

F-i1 = 46.939 kg/s
F-i2 = 46.939 kg/s
F-b1set = F-b2set = 23.168 kg/s
h-i1 = h-i2 = 6.9×10^5 J/kg

$F_{b1} = F_{b2} = 23.168$ kg/s
h = 6.9×10^5 J/kg
$F_f = 47.541$ kg/s

9 The Valve (Gas)

Inputs :
- x_pos = valve stem position (0 : 1)
- rho_ig = gas density [kg/m^3]
- p_ig = inlet gas pressure [Pa]
- T_ig = inlet gas temperature [°K]
- h_ig = inlet gas enthalpy [J/kg]

Outputs :
- w_g = outlet gas mass flow [kg/s]
- p_og = outlet gas pressure [Pa]
- rho_og = outlet gas density [°K]

- h_og = outlet gas enthalpy [J/kg]

Parameters :

- c_v2 = constant valve construction coefficient [m^2]
- k = ratio of specific heats for gas [-]
- delta_p = pressure drop across the valve [Pa]
- delta_p_rated = rated pressure drop across the valve [Pa]
- cp_ref = reference specific heat [J/(kg*°K)]

Steady state operating conditions :

Inputs :

X_{pos} = 0.95268

ρ_{ig} = 13.662 kg/m^3

p_{ig} = 45.251 kg/m^3

T_{ig} = 717.72 °K

h_{ig} = 3.3117 x10^6 J/kg

Outputs :

W_{eg} = 12 kg/s

P_{og} = 45.151x10^5 Pa

P_{og} = 13.685 kg/m^3

T_{og} = 717 °K

h_{og} = 3.3109x10^6 J/kg

Parameters :

c_v2 =0.053 m^2

k = 1.3

delta_p = 1x10^4 Pa

delta_p_rated = 1x10^4 Pa

cp_ref = 2.3x10^3 J/(kg*°K)

APPENDIX 2

POLYNOMIAL REPRESENTATION FOR SATURATED STEAM AND WATER CONDITIONS

A2.1 Introduction

Where-ever in the book saturated steam and water conditions are considered, the actual value of a state is approximated by a polynomial fit. The details are given below.

Nomenclature:

h_v- enthalpy of saturated steam
T_v - temperature of saturated steam
P_v - pressure of saturated steam
ρ_v - density of saturated steam
h_f - enthalpy of evaporation
h_w - enthalpy of saturated water
T_w - temperature of saturated water
P_w - pressure of saturated water
ρ_w - density of saturated water

A2.2 Saturated steam conditions

A2.2.1 Function $h_v(\rho_v) = f_6(\rho_v)$ (enthalpy as a function of density)

Appearance ; drum.f
Equations:

$$n_v = \ln(\rho_v)$$

$$h_v = a_{06} + a_{16}n_v + a_{26}n_v^2 + a_{36}n_v^3 + a_{46}n_v^4 + a_{56}n_v^5 + a_{66}n_v^6$$

where

a06 = 2700200.0
a16 = 47574.0
a26 =1117.6

a36=-526.3118
a56=-60.5790
a66 =-5.3885

A2.2.2 Function $T_V(\rho_V) = f_7(\rho_V)$ (temperature as a function of density)

Appearance : drum.f
Equations:

$$n_v = \ln(\rho_v)$$

$$T_V = a_{07} + a_{17} n_V + a_{27} n_V^2$$

where

a07 = 390.4075
a17 = 35.5266
a27 =2.7876

A2.2.3 Function $p_V(\rho_V) = f_8(\rho_V)$ (pressure as a function of density)

Appearance : drum.f
Equations:

$$p_V = a_{08} + a_{18} \rho_V + a_{28} \rho_V^2 + a_{36} n_V^3 + a_{38} \rho_V^4$$

where

a08 = 42819.0
a18 = 217030.0
a28 = -703.6933
a38= 0.7483

A2.2.4 Function $h_W(\rho_V) = f_9(\rho_V)$ (enthalpy of saturated water as a function of density of saturated steam)

Appearance : drum.f
Equations:

$$n_v = \ln(\rho_v)$$

$$h_W = a_{09} + a_{19} n_V + a_{29} n_V^2 + a_{39} n_V^3$$

where

a09 = 483140.0
a19 = 141310.0
a29 =16447.0
a39 =1373.1

A2.2.5 Function $h_f(\rho_v) = f_{44}(\rho_v)$ (enthalpy of evaporation as a function of pressure of saturated steam)

Appearance : superheater.f
Equations:

$$h_f = a_{044} + a_{144} p_v$$

where

a044 = 2.2464×10^6
a144 = 0.0892

A2.2.6 Function $h_v(T_v) = f_{11}(T_v)$ (enthalpy of steam as a function of temperature of steam)

Appearance : steam.f (condenser model)
Equations:

$$h_v = a_{011} + a_{111} T_v + a_{211} T_v^2 + a_{311} T_v^3 + a_{411} T_v^4$$

where

a011 = -1.8934×10^6
a111 = 4.1404×10^4
a211 = -148.7585
a311 = 0.2471
a411 = -1.5519×10^{-4}

Then, the derivative $\frac{dh_v}{dT_v}$ may be easily calculated

$$\frac{dh_v}{dT_v} = a_{111} + 2.0 a_{211} T_v + 3.0 a_{311} T_v^2 + 4.0 a_{411} T_v^3$$

A2.2.7 Function $\rho_v(T_v) = f_{12}(T_v)$ (density of steam as a function of temperature of steam)

Appearance : steam.f (condenser model)
Equations:

$$e_T = \exp\left(\frac{T_v}{75.0}\right)$$

$$\rho_v = a_{012} + a_{112} e_T + a_{212} e_T^2 + a_{312} e_T^3$$

where

a012 = -2.4494
a112 = 0.0222
a212 = -1.8195×10^{-6}
a312 = 8.9631×10^{-10}
a411 = -1.5519×10^{-4}

The derivative $\frac{d\rho_V}{dT_V}$ is calculated according to the equation

$$\frac{d\rho_V}{dT_V} = \frac{a_{111}}{75.0}e_T + \frac{2.0a_{211}}{75.0}e_T^2 + \frac{3.0a_{311}}{75.0}e_T^3$$

A2.2.8 Function $\rho_V(T_V) = f_{13}(T_V)$ (pressure of steam as a function of temperature of steam)

Appearance : steam.f (condenser model)
Equations:

$$e_T = \exp\left(\frac{T_V}{75.0}\right)$$

$$\rho_V = a_{013} + a_{113}e_T$$

where

a013 = $4.6772\text{x}10^5$
a113 = $4.1841\text{x}10^3$

The derivative $\frac{d\rho_V}{dT_V}$ is calculated according to the equation

$$\frac{d\rho_V}{dT_V} = \frac{a_{113}}{75.0}e_T$$

A2.3 Saturated water conditions

A2.3.1 Function $p_W(h_W) = f_3(h_W)$ (pressure of saturated water as a function of enthalpy of saturated water)

Appearance : drum.f , deaerator.f, economizer.f
Equations:

$$P_W = \begin{cases} a_{031} + a_{131}h_W + a_{231}h_W^2 + a_{331}h_W^3 & \text{if } h_W \le 567680 \\ a_{032} + a_{132}h_W + a_{232}h_W^2 & \text{if } h_W \le 567680 \end{cases}$$

where

a031 = 0.0
a131 = 0.1245
a231 = $-9.7369\text{x}10^{-7}$
a331 = $3.0143\text{x}10^{-12}$
a032 = $2.3815\text{x}10^6$
a132 = -10.1102
a232 = $1.0905\text{x}10^{-5}$

A2.3.2 Function $\rho_W(h_W) = f_4(h_W)$ (saturated water density as a function of saturated water enthalpy)

Appearance : drum.f
Equations:

$$\rho_W = a_{04} + a_{14} h_W + a_{24} h_W^2$$

where

a04 = 1003.4
a14 = -0.58372×10^{-4}
a24 = -1.1966×10^{-10}

A2.3.3 Function $T_W(h_W) = f_5(h_W)$ (saturated water temperature as a function of saturated water enthalpy)

Appearance : drum.f, economizer, f, deaerator.f
Equations:

$$T_W = a_{05} + a_{15} h_W + a_{25} h_W^2$$

where

a05 = 268.3632
a15 = 0.26922×10^{-3}
a25 = 0.34182×10^{-10}

A2.3.4 Function $h_W(T_V) = f_{14}(T_V)$ (enthalpy of saturated water as a function of temperature of saturated steam)

Appearance : steam.f (condenser model)
Equations:

$$h_W = a_{014} + a_{124} T_V$$

where

a014 = 1.1449×10^6
a114 = -1.1449×10^6

A2.3.5 Function $\rho_W(T_V) = f_{15}(T_V)$ (pressure of saturated water as a function of temperature of saturated steam)

Appearance : steam.f (condenser model)
Equations:

$$e_T = \exp\left(\frac{T_V}{75.0}\right)$$

$$\rho_W = a_{015} + a_{115} e_T + a_{215} e_T^2$$

where

a015 = 1.2630×10^4
a115 = -643.2748
a215 = 8.6655

This approximation is only valid for low temperature of steam (up to 393°K).

SUBJECT INDEX

Page numbers in bold typeface indicate a main subject reference.

MIX
Papier aus verantwortungsvollen Quellen
Paper from responsible sources
FSC® C105338
FSC
www.fsc.org

If you have any concerns about our products,
you can contact us on
ProductSafety@springernature.com

In case Publisher is established outside the EU,
the EU authorized representative is:
Springer Nature Customer Service Center GmbH
Europaplatz 3, 69115 Heidelberg, Germany

Printed by Libri Plureos GmbH
in Hamburg, Germany